D1545638

ALGAL PHOTO-SYNTHESIS

Current Phycology

Series editors: M. J. Dring, Queen's University Belfast
Michael Melkonian, Universität Köln

Also published

Ralph A. Lewin and Lanna Cheng, eds. *Prochloron: A Microbial Enigma*
Michael Melkonian, ed. *Algal Cell Motility*

ALGAL PHOTO-SYNTHESIS

Richard J. Geider
and Bruce A. Osborne

Chapman and Hall

New York

London

First published in 1992 by
Chapman and Hall
an imprint of
Routledge, Chapman & Hall, Inc.
29 West 35 Street
New York, NY 10001-2291

Published in Great Britain by
Chapman and Hall
2-6 Boundary Row
London SE1 8HN

Library of Congress Cataloging in Publication Data

Geider, Richard J., 1955–
    Algal photosynthesis : the measurement of algal gas exchange /
Richard J. Geider, Bruce A. Osborne.
        p.   cm.—(Current phycology)
    Includes bibliographical references and index.
    ISBN 0-412-02351-2
    1. Photosynthesis—Measurement. 2. Algal gas exchange—
—Measurement. 3. Algae—Physiology. I. Osborne, Bruce A., 1952–.
II. Title. III. Series.
QK565.G45   1991   91-7596
589.3'13342—dc20   CIP

British Library Cataloguing in Publication Data

Osborne, Bruce *1952–*
    Algal photosynthesis : the measurement of algal gas
    exchange.
    1. Algae. Photosynthesis
    1. Title II. Geider, Richard *1955–* III. Series
    581.13342
    ISBN 0-412-02351-2

To Clare, Kath, and our parents.

Contents

Preface

The algae are a diverse group of organisms with forms that range in size from less than a micron in diameter to over ten meters in length. Small (< 1 μm diam) unicellular forms dominate the open waters of the oceans and large lakes. Large multicellular forms often form a large component of the autotrophic biomass of shallow waters at the periphery of lakes and oceans but have also been found on seamounts in clear open ocean waters at depths up to several hundred meters (Littler and Littler, 1985). Phytoplankton in the sea probably account for more than 50% of global photosynthesis, although there is considerable uncertainty about this estimate. In addition, many symbiotic associations between unicellular algae and heterotrophic or autotrophic organisms have been identified, and algae can be found in a diverse range of terrestrial environments, ranging from polar to desert regions. The most important common biochemical attribute that unites the algae is their ability to split water, producing molecular oxygen during photosynthesis and concomitantly assimilating carbon dioxide. This attribute is shared with the terrestrial plants, cyanobacteria and chloroxybacteria. Although vascular plants are excluded from this review, we employ a broad definition of algae that includes the photosynthetic, oxygenic procaryotes.

Measurements of gas exchange are fundamental to most biochemical, physiological and ecological investigations of the algae. Information on the gas exchange characteristics of algae has progressed steadily over the past 80 years following the introduction of the unicellular green alga *Chlorella* to photosynthesis research by Otto Warburg. Already by the 1930's, gas exchange measurements carried out by Robert Emerson had led to such fundamental concepts as the photosynthetic unit. Adversaries for many years, Warburg and Emerson laid much of the groundwork for subsequent investigations in photosynthesis research.

Only under restricted conditions can photosynthetic reactions be completely separated from overall algal metabolism. For example, very short flashes of light ($< 1 \mu s$) have been used to investigate photosynthetic water splitting (oxygen evolution), and nutrient-starved "resting" cells have been used to investigate the biochemical pathway of carbon dioxide fixation (i.e., the Calvin or photosynthetic carbon reduction cycle) and the formation of simple sugars and starch. In general, however, gas exchanges due to photosynthetic reactions cannot be completely isolated from other metabolic gas exchanges and growth processes. Mitochondrial respiration proceeds at variable rates in illuminated cells, as do other oxygen-consuming and inorganic carbon-consuming and evolving reactions. An understanding of these other metabolic processes is necessary for interpreting measurements of gas exchange by algae.

The study of algal photosynthesis has progressed greatly during the last decade. Improvements and refinements in methodology have led to increases in our basic knowledge of photosynthetic and other gas exchange reactions. This information, however, is dispersed in the primary literature. In this book, we review the conceptual background, available methodology, and interpretation of the results of algal photosynthesis research. This review is not a technical manual but rather a summary and appraisal of the available methods for measuring photosynthetic gas exchange, including a discussion of some applications of these techniques. Our choice of subject matter covered reflects our background and research interests. We hope that there are no major omissions, although other investigations might have approached this review with a different emphasis.

In reviewing the methods available for measuring gas exchange by algae, we have included an overview of measurement techniques, an assessment of their sensitivity and accuracy, and a description of their limitations and applications. Wherever possible, the different techniques are compared, and potential explanations for any apparent differences are examined. Although algae can be separated from heterotrophs under controlled laboratory conditions, autotrophic and heterotrophic organisms coexist in the same environment in the real world. Determining the rate of photosynthesis by algal assemblages in nature requires corrections for the activity of heterotrophs, and experimental manipulations that allow these problems to be overcome are considered. Using fluorescence measurements to probe photosynthetic reactions is described also. In Chapters 1 and 2, we consider the problems encountered in manipulating organisms over the extreme size range, from less than 1 μm to more than 10 m, that is found in algae. Methods for determining the concentrations of photosynthetic pigments, the optical properties of algae, and the light field in which algae are suspended are also included, because these ancillary techniques are necessary for the interpretation of gas-exchange measurements. Despite the importance of artificial light sources in all aspects of algal photosyntheses research, there is little information available on the suitability of different lamps; therefore, chapter 4 considers the characteristics of these and other related accessories.

It is not sufficient simply to determine gas-exchange rates without placing the measurements into a theoretical framework. The basis of most physiological interpretations is the photosynthesis–light response (PI) curve, which describes the dependence of net gas exchange on irradiance. Various formulations of the PI curve are described, and the effect of environmental variables (e.g., light, nutrients, and temperature) on the light-saturated and light-limited rates of photosynthesis are summarized. Finally, the role of algae in global photosynthesis is briefly examined, together with methods for estimating algal biomass over wide geographic areas by remote sensing, using algorithms for estimating photosynthesis from biomass and environmental variables.

Although this book is aimed primarily at experimental phycologists, many of the techniques described are not specifically restricted to algal photosynthesis research and should also be of relevance to those working in plant physiology, biophysics, and biochemistry.

Acknowledgments

We would particularly like to thank Peg Conlon and Mary O'Brien for typing the manuscript. John Raven encouraged us to write the book and also reviewed the draft manuscript.

Chapter 1

Net Gas Exchange

Photosynthesis involves the coordinated operation of a series of biochemical and biophysical reactions starting with the absorption of photons and ending with the incorporation of inorganic carbon into stable organic compounds. All algae and the cyanobacteria, as well as terrestrial vascular plants, evolve oxygen during the light reactions that produce reductant ($NADP^+ \rightarrow NADPH$) and adenosine triphosphate (ATP). These products of the light reactions are subsequently consumed in CO_2 fixation by the photosynthetic carbon reduction cycle (PCRC). Competing with photosynthetic carbon reduction for NADPH and ATP are other processes, including photorespiration, nitrate reduction, mediated ion and gas transport between algal cell and environment, and various maintenance and synthetic processes. Although photosynthetic physiology can be conceptually isolated from the remainder of algal metabolism and in vitro separation of light and dark reactions of isolated chloroplasts has been accomplished (Trebst et al., 1958), in practice photosynthesis occurs simultaneously with, and cannot be separated from, the remainder of algal growth and maintenance processes. This complexity can lead to some confusion in the operational definition of the rate of photosynthesis.

The rate of photosynthesis can be defined as the rate of evolution of O_2 by the light reactions of photosynthesis or the rate of consumption of CO_2 and/or HCO_3^- by the dark reactions. Measurement of these rates is complicated by simultaneous O_2 consumption and CO_2 evolution, and, in practice, only net gas exchange is readily quantified. The use of tracers such as ^{18}O and ^{14}C allows investigation of rates of partial reactions, but may be subject to some uncertainty associated with isotope disequilibrium. This chapter considers methods for measuring net gas exchange, and the use of tracers is considered in Chapter 2.

Net gas exchange is the oldest and most widely employed procedure for measuring plant photosynthesis. It remains the standard against which other measurements are often compared. Changes in concentration of either dissolved oxygen (O_2) or total inorganic carbon (TCO_2) in an aqueous sample or CO_2 in a gas sample are used to determine net-gas-exchange rates. Limits on the magnitude of measurable rates are set by the concentration of O_2 and TCO_2 in water and the accuracy of methods for determining O_2 and TCO_2 concentrations. Our discussion begins with a consideration of these factors.

Factors Influencing the Distribution of O_2 and CO_2

The solubility of dissolved O_2 in water is determined by temperature, salt concentration, and atmospheric pressure (Hitchman, 1978; Benson and Krause, 1984). Over the range of conditions typical of most natural waters, the concentration of O_2 in equilibrium with the atmosphere varies by a factor of approximately three (Table 1.1), decreasing from 457 mmol·m^{-3} in distilled water at 0°C, to 163 mmol·m^{-3} at 40 parts per thousand (‰) salinity and 40°C (Benson and Krause, 1984). Wide variations from the thermodynamically defined equilibrium O_2 concentrations can result from biological and chemical oxygen-consuming or -producing reactions. These variations are often expressed in terms of percent saturation or apparent oxygen utilization (AOU).

Table 1.1.   Values for the equilibrium dissolved oxygen concentration mmol·m^{-3} at 1 atm pressure and various temperatures and salinities (after Benson and Kraus, 1984). For conversion from mmol·m^{-3} to mg·l^{-1} multiply by 3.19988 |x|10^{-2}.

Temperature (°C)	Salinity (‰)				
	0	10	20	30	40
0	457	426	397	370	345
10	353	331	310	291	273
29	284	268	253	238	224
30	236	224	212	200	190
40	200	190	181	172	163

Values calculated from:

$$\ln (O_2)_{eq} = -133.90205 + 1.575701 \times 10^5/T - 6.642308 \times 10^7/T^2$$
$$+ 1.243800 \times 10^{10}/T^3 - 8.621949 \times 10^{11}/T^4$$
$$- S(0.017674 - 10.754/T + 2140.7/T^2)$$

where T is the absolute temperature (in °K) and S is the salinity (in ‰). For more details and corrections to account for variations in atmospheric pressure from 1 atm see Benson and Krause (1984).

$$\% \text{ Saturation} = [O_2] / [O_2]_{eq} \qquad (1.1)$$

$$\text{AOU} = [O_2] - [O_2]_{eq} \qquad (1.2)$$

Where $[O_2]$ is the dissolved oxygen concentration $[O_2]_{eq}$ is the thermodynamically defined equilibrium dissolved oxygen concentration at the ambient temperature and salinity.

In upper ocean waters characterized by low algal abundance (< 1 µg chlorophyll a dm^{-3}) oxygen percent saturation ranges approximately from 90% to 110% with pronounced seasonal variations (Jenkins and Goldman, 1985; Spitzer and Jenkins, 1989; Shulenberger and Reid, 1981) and smaller diel variations (Tijssen, 1979). In eutrophic waters, the range in oxygen percent saturation can be much more pronounced (< 10 to $>$ 200%) over a diel cycle as a consequence of a higher biomass and greater rates of biological activity. The diel and seasonal ranges of dissolved oxygen concentration and percent saturation depend on the balance between biological activity, which is often limited by nutrient supply, advective and diffusive exchange between water masses, and exchange through the air–water interface. By correcting for exchange processes, it is possible to estimate rates of community photosynthesis and respiration either from diel measurements or from changes in oxygen concentration over longer time periods.

The concentration of dissolved CO_2 at air equilibrium in natural waters is much lower than that of O_2, ranging from approximately 25 mmol·m^{-3} at 0°C in freshwater, to 6 mmol·m^{-3} at 40°C in seawater of 40‰ salinity (Skirrow, 1975). However, air equilibrium is not always attained and CO_2 can be significantly enriched or depleted by the effects of biological activity on aqueous inorganic carbon equilibria. Ionized forms of inorganic carbon can be present in appreciable and often variable amounts depending on the pH and ionic content of the solution. Total inorganic carbon (TCO$_2$) is much more variable than dissolved CO_2 ranging from approximately 10 mmol·m^{-3} in acidic freshwater at 20°C, increasing to 2.3 mol·m^{-3} in seawater with a salinity of 35‰, and exceeding 100 mol·m^{-3} in bicarbonate-rich waters at high pH.

Dissolved inorganic carbon exists in a number of readily interconvertable forms (Stumm and Morgan, 1970; Skirrow, 1975). The most important reactions involved in this system are

$$CO_2 + H_2O \leftrightarrow H_2CO_3 \qquad (1.3)$$

$$H_2CO_3 \leftrightarrow HCO_3^- + H^+ \qquad (1.4)$$

$$HCO_3^- \leftrightarrow CO_3^{2-} + H^+ \qquad (1.5)$$

where the forms of inorganic carbon are aqueous carbon dioxide gas dissolved in water (CO_2), carbonic acid (H_2CO_3), bicarbonate ion (HCO_3^-) and carbonate ion (CO_3^{2-}). The total inorganic carbon concentration (TCO_2) is the sum of all of these forms:

$$TCO_2 = [CO_2] + [H_2CO_3] + [HCO_3^-] + [CO_3^{2-}] \qquad (1.6)$$

Because of the difficulty in separately measuring the equilibria for Eqs. 1.3 and 1.4, these equations are often combined.

$$CO_2 + H_2O \leftrightarrow HCO_3^- + H^+ \qquad (1.7)$$

Of the different forms of inorganic carbon present in water, only CO_2 and TCO_2 are amenable to direct experimental observation. The concentrations of the other forms must be calculated from thermodynamic considerations. The equilibrium inorganic carbon speciation is influenced by coexisting equilibria involving the ionization of water ($H_2O \leftrightarrow H^+ + OH^-$), ionization of weak acids such as boric acid ($H_3BO_3 + H_2O \leftrightarrow B(OH)_4^- + H^+$), hydrogen sulfide ($H_2S \leftrightarrow HS^- + H^+$), or ammonium ($NH_3 + H_2O \leftrightarrow NH_4^+ + OH^-$), and the formation of ion pairs with various cations (Na, K, Ca, Mg) or the hydrolysis of aluminum ions ($Al^3 + 2H_2O \leftrightarrow Al(OH)^{2+} + 2H^+$). The most important physical and chemical properties used for describing the equilibrium concentrations of various inorganic carbon species include temperature, pH, and carbonate alkalinity.

Alkalinity (equivalents dm^{-3}) is defined as the sum of the normalities (i.e., the product of charge and molarity) of the anions of weak acids. In most waters, the dominant weak acids are those of the inorganic carbon system, and total alkalinity (TA) is defined as

$$TA = [HCO_3^-] + 2[CO_3^{2-}] + [OH^-] - [H^+] + c_{sa} \qquad (1.8)$$

where c_{sa} is the surplus alkalinity due to anions of weak acids other than carbonic acids. Carbonate alkalinity is the contribution of the first two terms on the left-hand side of Eq. 1.8. The contributions of $[OH^-]$ and $[H^+]$ will be significant under acidic or basic conditions, but can be ignored in most natural waters at a normal pH of 6–8. In sea waters, the contribution of borate to total alkalinity cannot be neglected, but a correction based on salinity can be applied (Skirrow, 1975). Organic acids may comprise a variable, but substantial, portion of total alkalinity in freshwaters (Herczeg and Hesslein, 1984; Herczeg et al., 1985) and have been suggested to comprise a small but measurable component of alkalinity in seawater (Bradshaw and Brewer, 1988a, b). The contributions of

H_2S and NH_4^+ to total alkalinity are usually unimportant under aerobic conditions where photosynthetic gas exchange is being measured, but may be important in suboxic or anoxic waters.

Oxygen Determinations

The concentration of dissolved O_2 can be measured by a number of techniques. Many early studies employed manometry, which is still advocated by some researchers (Dawes, 1985). However, most recent investigations have been undertaken with the Winkler method or with polarographic oxygen electrodes. Alternative methods, which were used in a limited number of investigations, but will not be considered further here, include gas chromatography for the simultaneous determination of O_2 and other gases (Weiss and Craig, 1973), microgasometry for the determination of O_2 in waterlogged soils and other fluids containing high concentrations of organic matter or solids (Scholander et al., 1955), and mass spectrometry (Benson and Parker, 1961).

Winkler Method

The Winkler method (1888) is the oldest method used to quantitatively determine dissolved O_2. It is based on the quantitative oxidation of Mn(II) to Mn(III) by O_2 in an alkaline solution followed by the oxidation of I^- to I_2 in acid solution and titration of I_2 with thiosulfate. This *fixation* of O_2, which comprises the initial steps of the Winkler technique, is summarized by the following reactions (Hitchman, 1978; Grasshoff, 1981):

$$Mn^{2+} + 2OH^- \rightarrow Mn(OH)_2 \tag{1.9}$$

$$2Mn(OH)_2 + 0.5O_2 + H_2O \rightarrow 2Mn(OH)_3 \tag{1.10}$$

$$2Mn(OH)_3 + 6H^+ + 2I^- \rightarrow 2Mn^{2+} + I_2 + 6H_2O \tag{1.11}$$

$$I_2 + I^- \leftrightarrow I_3^- \tag{1.12}$$

Once these reactions have been completed, a sample can be allowed to stand for a period of up to one day before completing the analysis. In the Winkler method, the O_2 content of a sample is determined without recourse to an O_2 standard. Instead, the technique relies on the quantitative stoichiometric conversion of O_2 to I_2 following by measurement of the amount of I_2 produced by titration with thiosulfate ($S_2O_3^{2-}$) to the starch endpoint, although potentiometric or amperometric endpoints can also be employed (Hitchman, 1978; Grasshoff, 1981). The strength of the thiosulfate titrant is normally determined by titrating a standard potassium iodate solution.

$$I_3^- + 2S_2O_3^{2-} \rightarrow 3I^+ + S_4O_6^{2-} \tag{1.13}$$

Spectrophotometric measurements of I_2 at 287.5 nm (Parsons et al., 1984) or I_3^- at 352 nm (Broenkow and Cline, 1969), although less accurate than carefully conducted iodometric titrations, provide a quick method for determining O_2 content to two significant figures.

Modifications to the original Winkler method by Montgomery et al. (1964) and Carpenter (1965b) have resulted in significant improvements over the original formulation, and the accuracy of the modified Winkler method has been experimentally demonstrated (Carpenter, 1965a). Descriptions of the modified Winkler technique, together with practical advice on its application are given by Hitchman (1978) and Grasshoff et al. (1983). Currently accepted protocols described by the American Public Health Association (1985) are not as precise as the protocols described by Montgomery et al. (1964) and Carpenter (1965b). Recent improvements in methodology involve automation of the titration using photometric (Williams and Jenkinson, 1982) or amperometric (Culberson and Huang, 1987) endpoints. A high accuracy (standard deviation of 0.3 mmol·m^{-3}) can be obtained with the Winkler technique provided that precise control is exercised over sample and reagent volumes.

Samples for O_2 analysis are usually collected in 50–300 cm^3 glass bottles with tapered ground-glass stoppers. Entrainment of atmospheric O_2 can be a major source of error in the technique and must be avoided during sample collection. Bottles are usually filled from a reservoir, or water bottle sampler, by inserting a tube to the bottom of the O_2 bottle and allowing liquid to overflow for at least three bottle volumes before addition of the Winkler reagents (American Public Health Association, 1985). Alternatively, smaller samples can be collected in 5.0–50 cm^3 glass syringes with tight-fitting ground-glass plungers, provided that care is taken to avoid atmospheric contamination and cavitation.

Loss of iodine can be a significant source of error in the Winkler technique, and a number of precautions can be taken to obtain the highest accuracy (Carpenter, 1965b; Grasshoff et al., 1983). These include protecting samples from bright light to prevent the photooxidative consumption of I_2 after acidification, and titrating the entire sample in the original bottle (rather than removing a subsample for titration), to avoid volatilization of I_2. Other problems can be encountered with the Winkler technique in waters containing high concentrations of organic matter, nitrite, ferrous iron, or sulfites, and modifications to the recommended procedures (Montgomery et al., 1964; Carpenter, 1965b) have been developed to overcome these potential problems (Hitchman, 1978; American Public Health Association, 1985).

Oxygen Electrodes

Oxygen electrodes are most often used in investigations with laboratory cultures of microalgae or thalli of macroalgae in which the ratio of plant biomass to volume of suspending medium is high. Polarographic O_2 electrodes were introduced to photosynthesis research in the late 1930s, when a hanging-drop mercury electrode was used in studies of oxygen evolution and measurements of the photon yield of photosynthesis in *Chlorella* (Petering et al., 1939). Grasshoff (1981) describes the application of hanging-drop mercury electrodes for measuring O_2 with specific applications to oceanographic research, although the potentially toxic effects of mercury may limit the usefulness of this method for continuous measurement of oxygen exchange.

Fixed platinum electrodes (Grasshoff, 1981) were used to obtain qualitative measurements of photosynthesis in vascular plant leaves and algal thallus tissue by Blinks and Skow (1938) and Haxo and Blinks (1950), although reproducibility in these early studies was poor. Olson et al. (1949) measured the transient current associated with an alternating potential applied to the electrode to obtain continuous measurements of O_2 concentration with a response time of seconds. The method was used to measure the photon yield of O_2 evolution by *Chlorella pyrenoidosa* (Brackett et al., 1953a, b). The rapid response of bare platinum electrodes make them especially suitable for specialized measurements, such as the determination of photosynthetic action spectra (Haxo and Blinks, 1950; Haxo, 1985), oxygen flash yields (Ley and Mauzerall, 1982), and measurements of S states* (Joliot and Joliot, 1968) but these electrodes may be subject to considerable error (see Meunier & Popovic, 1988).

The most commonly employed oxygen electrode for routine use is the membrane-covered polarographic detector (Hitchman, 1978; Grasshoff, 1981). This type of electrode, known as the Clark electrode, was designed to overcome problems associated with drift in calibration due to changes in the chemical composition of the sample and poisoning of the electrode from surface-active components that may contaminate bare electrodes (see above). The Clark type electrode consists of a platinum cathode, silver anode, and KCl salt solution, which are separated from the sample

---

*The S state refers to a state of reduction of the water oxidizing (i.e., oxygen evolving) component of photosystem II.

by a gas-permeable membrane. According to Grasshoff (1981) the reaction of the cathode ($O_2 + 4H^+ + 4e^- \rightarrow 2 (H_2O)$) is coupled to the reaction at the anode ($4Ag \rightarrow 4Ag^+ + 4e^-$) by the electrolyte KCl ($Ag^+ + Cl^- \rightarrow AgCl$).

Membrane-covered electrodes are produced by a number of manufacturers and are routinely used in many laboratories. They are less subject to drift or contamination than bare platinum electrodes and allow much more rapid measurement of O_2 concentration than the Winkler method. These electrodes, typically, exhibit a 90% response time of 30 sec and a precision of 10 mmol·m^{-3} (Hitchman, 1978). The response time is considerably slower than that obtained with a bare platinum electrode, and the precision is 10 to 100 times less than can be obtained from the Winkler titration. Use of the membrane-covered electrode requires that the sample be stirred to break down boundary layers in the aqueous phase so that diffusion through the membrane becomes the step that limits O_2 supply to the electrode surface.

The output from membrane-covered electrodes is linearly related to the partial pressure of O_2 (pO_2) in the solution. To obtain absolute O_2 concentration requires additional information on temperature, salinity, and atmospheric pressure so that pO_2 can be converted to a concentration. Temperature must be carefully controlled, because electrode properties and response, particularly the diffusion of O_2 through the membrane, are highly temperature dependent (Hitchman, 1978). Routine calibration of commercial electrodes at two points, usually oxygen-free water (a solution of 2% sodium sulfite or a supply of oxygen-free nitrogen gas) and air-saturated water, is recommended (Hitchman, 1978). Electrodes can also be calibrated in water-saturated air, however, care must be taken to ensure that the electrode surface is free of contaminants and that the air is completely saturated with water vapor.

Polarographic O_2 *micro*electrodes are becoming a standard component of some specialized areas of photosynthesis research. The construction and use of these electrodes is described by Revsbech and Jorgensen (1986). Electrodes with a tip diameter of only a few micrometers are useful for photosynthesis measurements that require high spatial resolution. These electrodes have been used to measure photosynthesis within benthic algal mats, by symbiotic algae associated with planktonic foraminifera, or by algae within macroscopic aggregates. They may also prove to be useful in investigations of oxygen fluxes and concentration within the diffusive boundary layer around macroalgae. Despite the expectation that diffusion through a boundary layer at the electrode surface should be negligible for microelectrodes, Gust et al. (1987) report that the response of microelectrodes in still water is only 10% to 50% of the maximum response in stirred water, which may lead to artifacts

Table 1.2.   Typical reported precision of various techniques for measurements of dissolved gas concentrations.

Technique	Precision	Reference
O_2 electrode-steady mode	3–6 mmol·m^{-3}	Hitchman, 1978
O_2 electrode-pulsed mode	0.8 mmol·m^{-3}	Langdon, 1984; Bender et al. 1987
Winkler titration-visual detection	0.7 mmol·m^{-3}	Strickland, 1960 Talling, 1976
Winkler titration-automated detection	0.3 mmol·m^{-3}	Williams and Jenkinson, 1982, Culberson and Huang, 1987
Total CO_2 from pH and aklalinity	5 mmol·m^{-3}	Smith and Key, 1975
Total CO_2 from potentiometric titration	2 mmol·m^{-3}	Bradshaw and Brewer, 1988a,b
Total CO_2 from acidification and IRGA[a]	<3–6 mmol·m^{-3}	Johnson et al., 1981 Roberts and Smith, 1988
Total CO_2 from coulometric titration	1 mmol·m^{-3}	Bender et al., 1987 Johnson et al., 1987

[a]Infrared gas analysis.

when measuring O_2 concentrations in gradients of water velocity such as occur at solid surfaces in flowing waters.

A computerized, pulsed membrane-covered electrode with a precision of 0.8 mmol·m^{-3} approaches the sensitivity of the Winkler technique (Langdon, 1984). This approach relies on measuring the transient current associated with application of a pulsed potential across the electrochemical cell (Hitchman, 1978). The electrode does not require stirring and is thus suitable for use with fragile microalgae such as the dinoflagellate *Gonyaulax tamarensis* (Langdon, 1978), which may be damaged by conventional technique. An instrument based on the pulsed electrode can be deployed for periods of 3–4 months in the sea (Maccio and Langdon, 1988) to provide continuous monitoring of oxygen in situ.

Direct comparisons of measurements of O_2 obtained with the Winkler technique and membrane-covered polarographic electrodes generally indicate good agreement within the limits set by the precision of the analytic techniques (Hitchman, 1978; Langdon, 1984). When sufficient care is taken, the Winkler technique provides higher precision than the O_2 electrodes (Table 1.2); but, because it is destructive, the Winkler technique allows only one O_2 determination per sample. Therefore, at least two samples are required for each rate determination. The Winkler method is used primarily in field studies when a large number of samples are incubated for an extended period of time (usually 1–24 hr). The high precision and accuracy of the technique, together with its ability to rapidly fix the measurable oxygen concentration in a large number of samples for subsequent analysis makes this technique suitable for routine use in the field. Conventional oxygen electrodes are usually unsuitable for studies of phytoplankton photosynthesis, especially in oligotrophic waters. However, Maccio and Langdon's (1988) pulsed electrode has been used to monitor diel changes (O_2) in open ocean waters.

Inorganic Carbon Determinations

Total inorganic carbon can be measured directly by acidifying a sample and determining the amount of liberated CO_2 by infrared gas analysis (Roberts and Smith, 1988), by gas chromatography, or coulometric titration (Johnson et al., 1985; 1987). These techniques have an accuracy of approximately 1–5 mmol·m^{-3} TCO_2 in ocean waters where the TCO_2 concentration is typically 2 mol·m^{-3}. Thus, on a molar basis, the best TCO_2 measurements are approximately a factor of three less sensitive than the best O_2 determinations. Direct determination of net TCO_2 fluxes have been undertaken in a limited number of investigations (Hofslagare et al., 1985).

Alternatively, TCO_2 can be calculated from observations of pH and alkalinity (Smith and Kinsey, 1978; Hofslagare et al., 1985), based on a thermodynamic treatment of the inorganic carbon dioxide system (Skirrow, 1975). Under the assumption that the inorganic carbon system is in equilibrium, total inorganic carbon is given by

$$[TCO_2] = [CA] \frac{1 + K_2^1/H^+ + H^+/K_{L1}^1}{1 + 2K_2^1/H^+} \tag{1.14}$$

where $[TCO_2]$ is the total inorganic carbon concentration, $[CA]$ is the carbonate alkalinity, $[H^+]$ is the hydrogen ion activity (i.e., 10^{-pH}), K_{L1}^1 is the first apparent dissociation constant for carbonic acid and K_2^1 is the second apparent dissociation constant of carbonic acid in water (values for these constants are available in Skirrow, 1975).

Photosynthetic CO_2 fixation produces an increase in the pH of the suspending medium without altering the alkalinity. Thus, changes in pH together with an initial value of the carbonate alkalinity, can be used to continuously assess changes of net inorganic carbon concentration using Eq. 1.14. The concentration of TCO_2 in seawater can be measured to an accuracy of ±5 mmol·m^{-3} when pH is measured with an accuracy of ±0.005 units and alkalinity is measured to ±5 mmol·m^{-3} (Smith and Key, 1975). This is the practical limit on measurement of TCO_2 in discrete samples. However, measurement of a much smaller change in TCO_2 concentration of only ±0.5 mmol·m^{-3} is possible assuming that changes in pH of 0.0005 units can be detected by continuously monitoring pH in a single sample (Axelsson, 1988). Very accurate determinations of alkalinity (± 1 mmol·m^{-3}) and TCO$_2$ (± 1 mmol·m^{-3}) in sea water by potentiometric titration requires stepwise addition of acid and application of elaborate algorithms to calculate the equivalence points on the titration curve for H^+, HCO_3^-, HSO_4^-, HF, $B(OH)_4^-$, $H_2PO_4^-$, and $SiO(OH)_3^-$ (Bradshaw et al., 1981).

TCO$_2$ concentrations can also be calculated from measurements of pH and aqueous CO$_2$ [CO$_2$(aq)] (Tregunna and Thomas, 1968; Browse, 1985) under the assumption that CO$_2$ in the gas and aqueous phases are in equilibrium:

$$[TCO_2] = K_H[CO_2(g)]\left(1 + \frac{K_{L1}^1}{H^+} + \frac{K_{L1}^1 K_2^1}{(H^+)^2}\right) \qquad (1.15)$$

where CO$_2$(g) is the concentration of CO$_2$ in the gas phase that is in equilibrium with the aqueous phase, and K_H is Henry's law constant.

Inaccuracies in measuring pH due to differences in the ionic strength of the sample and buffers used to calibrate the pH meter may bias calculated rates of TCO$_2$ exchange. Axelsson (1988) showed that calculated rates of net photosynthesis of marine macroalgae were 20% greater when the pH meter was standardized against the National Bureau of Standards (NBS) pH scale than when the meter was calibrated against the Hansson (1973) seawater pH scale. Direct determination of the TCO$_2$ changes associated with the observed changes in pH were intermediate between those obtained from the indirect calculations using the NBS and Hansson scales (Axelsson, 1988). Difficulties in measuring pH may be exacerbated in waters with low ionic strength (McQuaker et al., 1983), particularly in the presence of high concentrations of dissolved organic carbon (Herczeg and Hesslein, 1984). In fact, it has been suggested that pH in some low alkalinity freshwaters containing high concentrations of dissolved organic carbon should be calculated from the more accurate measurements of TCO$_2$ and the partial pressure of CO$_2$ to overcome biases in pH measurements (Herczeg et al., 1985).

Changes in pH and alkalinity brought about by metabolic processes other than inorganic carbon exchange can lead to inaccuracies in estimates of TCO$_2$ exchange if constant alkalinity is assumed. The consumption or production of H^+ or OH^- accompanies many metabolic processes (Smith and Raven, 1976), as indicated in Table 1.3. These include assimilation of inorganic nitrogen, calcium carbonate precipitation, and organic acid excretion. The potential effects of these processes on estimates of TCO$_2$ exchange are now considered.

Table 1.3.   Summary of major processes leading to extracellular pH changes.

Metabolic Process	Direction of External pH Change
Photosynthetic CO_2 fixation	Increase
Nitrogen assimilation: NH_4^+ entering cell	Decrease
Nitrogen assimilation: NO_3^- entering cell	Increase
Excess cation influx, e.g., K^+ carboxylate accumulation	Decrease
Excess anion flux, e.g., Cl^-/HCO^- exchange	Increase
Organic acid assimilation— undissociated form entering	Increase
Excretion of organic acids in undissociated form	Decrease
Calcium carbonate precipitation	Increase

Adapted from Smith and Raven (1976).

The typical stoichiometry of inorganic carbon and nitrogen assimilation by algae and the associated oxygen and proton fluxes is given by Eqs. 1.16–1.18

$$106CO_2 + 138H_2O + 16NO_3^- \Rightarrow \tag{1.16}$$
$$(CH_2O)_{106}(NH_3)_{16} + 16OH^- + 138O_2$$

$$106CO_2 + 106H_2O + 16NH_4^+ \Rightarrow \tag{1.17}$$
$$(CH_2O)_{106}(NH_3)_{16} + 16H^+ + 106O_2$$

$$106CO_2 + 114H_2O + 8CO(NH_2)_2 \Rightarrow \tag{1.18}$$
$$(CH_2O)_{106}(NH_3)_{16} + 8CO_2 + 106O_2$$

Note the one-to-one relationships between NO_3^- and OH^- in Eq. 1.16 and between NH_4^+ and H^+ in Eq. 1.17. Goldman and Brewer (1980) verified these stoichiometries for nitrogen assimilation and H^+ or OH^- exchange in experiments with the marine chlorophyte *Dunaliella tertiolecta*. They demonstrated a one-to-one relationship between nitrogen assimilation and alkalinity changes when NH_3^+, NO_2^-, or NO_3^- were the nitrogen sources. Assimilation of the cation ammonium decreased alkalinity, whereas assimilation of the nitrate or nitrite anions increased alkalinity. In contrast, assimilation of urea does not produce a net acid or base exchange, and alkalinity remained constant when phytoplankton used urea as a nitrogen source.

Warm ocean waters and many hard-water lakes are supersaturated with respect to calcium carbonate, and $CaCO_3$ precipitation can occur by biological or chemical processes (Broeker and Peng. 1982), that is,

$$Ca^{2+} + 2HCO_3^- \Rightarrow CaCO_3(s) + H_2CO_3 \qquad (1.19)$$

where s refers to the solid phase. Calcium carbonate precipitation reduces alkalinity, with a stoichiometry of 0.5 $mol \cdot m^{-3}$ $CaCO_3$ formed producing a 1.0 $mol \cdot m^{-3}$ reduction in total alkalinity (Smith and Key, 1975). A wide variety of algae can precipitate $CaCO_3$ including the coraline algae, which are important contributors to reef ecosystems and the coccolithophorids, which are widespread in the phytoplankton (Ackleson et al., 1988), as well as the microalgae associated with foraminifera and corals (Borowitzka, 1977; Pentecost, 1985). However, calcium carbonate precipitation only occurs in waters that are nearly saturated or supersaturated with calcite or aragonite (Pentecost, 1985).

Calcification rates covaried with photosynthesis rates in four *Halimeda* species, with calcium carbonate precipitation proceeding at approximately 64% of the rate of organic carbon production (Jensen et al., 1985). The rate of calcification was found to equal the photosynthesis rate in *Corallina officinalis* (Pentecost, 1978), and the coccolithophorid *Emiliania huxleyi* (Paasche, 1964), but lower ratios of < 0.25 were obtained for *Halimeda* spp. (Borowitzka and Larkum, 1976). Both photosynthetic oxygen evolution and calcification rates could be described as exponential or hyperbolic tangent functions of irradiance in the reef-building corals *Manicura areolata*, *Acropora cervicornis*, and *Acropora formosa*. From the limited number of samples examined, both photosynthesis and calcification rates appeared to saturate at similar irradiances (Chalker, 1981).

Organic acids are important early products of photosynthesis and photorespiration in algae, but the importance of these compounds in altering pH during photosynthesis experiments is expected to be minimal under conditions in which low-molecular-weight organic acids usually make up $< 10\%$ of the exuded dissolved organic carbon (DOC). As DOC release is typically $< 10\%$ of gross photosynthetic carbon fixation this would not seem to be a problem. However, the presence of organic acids in fresh and seawaters may affect the determination of carbonate alkalinity and thus calculated TCO_2 concentrations. The error is not large in ocean waters where a careful comparison of indirect and direct estimates of TCO_2 concentration (Bradshaw and Brewer, 1988a, b) demonstrated that indirect estimates of TCO_2 calculated from high-precision measurements of alkalinity exceeded direct measurements of TCO_2 by approximately 1% (21 $\mu mol \cdot kg^{-1}$). This difference between the two methods was attributed to an unknown protolyte, probably the anion of a weak organic acid with humic and fulvic acids suggested as likely candidates.

Measuring CO_2 Exchange in Air

Some of the difficulties in measuring carbon exchange by algae can be overcome if plants are suspended in a moist atmosphere that allows rapid and direct measurement of CO_2 fluxes by infrared gas analysis (IRGA). Bidwell and McLachlan (1985) found that CO_2 assimilation rates were stable for periods of up to 8 hours, and that both photosynthesis and respiration proceeded at similar rates whether plants were suspended in air or water. Addition of carbonic anhydrase to the plant surface was necessary to insure rapid intercoversion of CO_2 and HCO_3^- since HCO_3^- was actively assimilated (Bidwell and McLachlin, 1985). A similar technique can be employed with microalgae, in which cells at a density of 2 mg chlorophyll·dm^{-2} are supported in a thin film of water on a nylon screen (Bidwell, 1977). Bidwell (1977) reported steady rates of net photosynthetic or respiratory gas exchange of *Chlorella pyrenoidosa* for periods of up to 15 hr provided that incoming air was saturated with water and found similar rates in suspended cultures containing carbonic anhydrase and artificial leaves. More recently, Vonshak et al. (1989) have measured O_2 exchange rates of cells of the cyanobacterium *Spirulina platensis* suspended on glass-fiber filters. However, the extensive handling required to prepare an artificial leaf may make this technique inappropriate for many fragile species lacking a rigid cell wall.

Maintaining a high tissue water content is essential to the maintenance of photosynthesis rates of macroalgae in air. Photosynthesis by *Ulva* was not affected between the onset of desiccation and up to 20% water loss, but net photosynthesis was completely inhibited when plant water loss equalled 50–70% (Beer and Eshel, 1983). However, even a 10% water loss resulted in approximately a fourfold reduction in carbon fixation rates with *Enteromorpha prolifera* (Pregnall, 1983). In general, algae from the lower intertidial zone show reduced rates of photosynthesis in air when compared with submerged rates, whereas middle and upper intertidal algae appear to photosynthesize more rapidly in air than in water (Johnson et al., 1974; Brinkhaus et al., 1976; Quadir et al., 1979).

Measuring Net Gas Exchange

The simplest system for measuring net gas exchange consists of a closed container in which O_2 or TCO_2 is measured at the start and end of an incubation. The first measurements of photosynthesis in aquatic environments were obtained with such a method by using the Winkler technique to follow changes of dissolved oxygen concentration in illuminated and darkened bottles (Gaarder and Gran, 1927). This method consists of dispensing subsamples containing a plant population into clear and opaque glass bottles, measuring the initial dissolved oxygen concentration in one set of replicate bottles and measuring the final concentrations in sets of light and dark bottles. Oxygen evolution in the light is referred to as net productivity, and oxygen consumption in the dark provides a measure of respiration. It is commonly assumed that respiration in the dark bottle occurs at the same rate as respiration in the light bottle, and the difference between oxygen concentrations in the light and dark bottles at the end of the experiment is assumed to equal total photosynthetic oxygen evolution, termed gross productivity. The assumption that the rates of O_2 consumption in dark and light bottles are equal is a major limitation of the technique for estimating "true" gross photosynthesis because of the possibility of light-dependent O_2-consuming processes and modifications of mitochondrial respiration by light (see Chapter 2). The light–dark bottle technique, however, has been widely and repeatedly used in aquatic ecosystems (Strickland, 1960), and it remains a standard technique in limnological and oceanographic research on algal productivity.

The light–dark bottle oxygen technique was used in an extensive series of oceanic investigations by Riley (1938, 1939, 1941, 1944) which provided the first estimate of gross photosynthesis rates in the open ocean of 860 ± 560 mg C·m^{-2}·d^{-1} (range of 165–1,550 mg C·m^{-2}·d^{-1}). In some regions, incubations of up to 7 d duration were required to obtain measurable differences in oxygen concentration between light and dark bottles. Riley's experiments in oligotrophic waters were later criticized by Steemann-Nielsen (1952) because of potential artifacts associated with the long incubation times. Steemann-Nielsen found a much lower rate of primary production with his new ^{14}C technique. Improvements in methodology now allow measurements of O_2 production and consumption rates in incubations that last less than a day even in the most oligotrophic open ocean waters (Williams et al., 1983). Rates of gross photosynthesis in the oligotrophic (< 0.1 μg chlorophyll a·dm^{-3}) subtropical North Pacific gyre averaged 1.3 μmol O_2·dm^{-3}·d^{-1} in the upper mixed layer (Williams et al., 1983). Assuming that this rate applies to the upper 50 m of the water column, and that the photosynthetic quotient is unity, an estimate of the areal rate of 780 mg C·m^{-2}·d^{-1} can be obtained. Although subject to considerable uncertainty, Riley's (1941) earlier conclusions appear to be consistent with this extrapolation of the results of Williams et al. (1983).

Providing sufficient biomass to ensure measurable changes in O_2 or

Table 1.4.   Effect of water movement on rates of photosynthetic and respiratory gas exchange in macroalgae.

| Macroalgae | Ratio of Rates in Mixed versus Stagnant Containers | | |
	Photosynthesis	Respiration	Reference
Ulva lobata	3.6	1.3	Littler (1979)
Colpomenia sinuosa	1.1	0.8	Littler (1979)
Padina sp.	2.6	—	UNESCO (1973)
	1.1	1.3	Pfeifer and
Cladophora sp.			McDiffett (1975)
	—	1.7	Whitford and
Oedogonium kurzii			Schumacher (1961)

Adapted from Littler (1979).

TCO_2 concentration does not pose a problem in measuring net gas exchange of macroalgae, and most investigators prefer to use O_2 electrodes rather than the Winkler titration to measure changes in O_2 concentration which generally exceed 10% of the initial value. In fact, care must be taken to ensure that optimum ratios of thallus weight to bottle volume are employed, because high ratios may lead to shading within the incubation bottle and/or rapid depletion of nutrients, including inorganic carbon, with affects on subsequent rates of photosynthesis during a prolonged incubation (Littler, 1979). Littler (1979) recommends a maximum dry weight to volume ratio of 0.03 $g \cdot dm^{-3}$ for experiments of up to 6 hr duration, with the further provision that whenever feasible "the proportions of the container be commensurate with the size and metabolic rate of an entire representative thallus." The duration of the incubation also influences the choice of container volume and sample biomass, with shorter incubations requiring correspondingly higher biomass. Water movement is greatly reduced when algae are enclosed in bottles, and it is necessary to provide mechanical mixing to avoid rate limitations due to diffusion effects (Kanwisher, 1966; Littler, 1979). Stirring appears to have more of an effect on photosynthesis than respiration (Table 1.4).

Oxygen and pH electrodes can be used to continuously measure O_2 or TCO_2 concentrations in either closed or open systems. When employed with closed systems under nearly constant environmental conditions, the rate of net gas exchange can be directly measured from the rate of O_2 or pH change. A constant rate of net gas exchange is usually established within minutes following a change of environmental conditions. Small chambers of approximately 1–10 cm^3 are often employed to measure net gas exchange of microalgae in the laboratory. Commercially available chambers are often illuminated from the side through a curved surface making determinations of irradiance within the experimental chamber difficult. Construction of a chamber that allows illumination through an optically flat surface to provide a well-defined light field is described by Dubinsky et al. (1987). A leaf disc electrode developed for use with terrestrial vascular plants (Delieu and Walker, 1981) can be employed to measure O_2 exchange by discs cut from a macroalgal thallus; however, wound respiration may be a problem with such measurements (Bidwell and McLachlan, 1985). Larger chambers can be used to measure rates of gas exchange of entire plants.

The change of O_2 and pH brought about by algal photosynthesis in closed chambers can influence the subsequent gas exchange rates. In some circumstances the large excursions of O_2 and pH that are possible in closed systems provide the data desired by an investigator. For example, the upper limit of pH at which net gas exchange ceases can be used to infer the chemical forms of inorganic carbon used for photosynthesis (Maberly and Spence, 1983; Axelsson and Uusitalo, 1988), and the effect of pO_2 on net gas exchange provides information on the rate of photorespiratory metabolism in algae (Burris, 1981). However, in many cases, large excursions of O_2 or pH are an undesirable feature of a closed system and must be compensated for by periodic renewal of TCO_2 or removal of O_2. As a practical limit, many investigators attempt to control pO_2 to within 80% to 120% saturation in order to limit O_2-sensitive changes in gas exchange rates when determining the light or temperature dependence of photosynthesis. Exchange of O_2 between gas and liquid phases is relatively rapid, and intermittent aeration can be used to maintain pO_2 near air saturation (Jassby, 1978). However, equilibration of pH and TCO_2 by aeration is much slower. Harris and Lott (1973) describe a system for measuring photosynthesis of freshwater phytoplankton where pH and O_2 are controlled by titrating with CO_2 and bubbling with N_2.

Open systems have been used to provide improved control of the environmental conditions within an experimental chamber. In systems that are open to fluid flow but closed to the atmosphere, measurements of the change of O_2 or TCO_2 between the entrance and exit ports of the chamber can be used to obtain gas exchange rates when the volume of the chamber and the flow rate through the chamber are known (Axelsson, 1988). Chambers that are open to the atmosphere but closed to fluid flow are also used in algal photosynthesis research. An IRGA can be used to continuously monitor TCO_2 exchange provided that aeration is vigorous enough to overcome rate limitation by CO_2 equilibration between gas and liquid phases (Browse, 1979). Measuring changes in O_2 of a gas stream used to flush a liquid sample is impractical because of the high concentration of O_2 in the atmosphere. However, differences in pO_2 between aqueous and gas phases can be used to estimate net gas exchange, provided that the kinetics of gas transfer through the air–water interface have been adequately characterized (Gallegos et al., 1980, 1983).

Sampling and Subsampling

The organisms under consideration here range over seven orders of magnitude in size from the < 1 μm diam chrococcoid cyanobacteria, chloroxybacteria, and some eucaryotic microalgae, to the > 10 m length of the giant kelp *Macrocystis pyrifera*. Various problems arise in attempts to manipulate organisms over this size continuum. Gas exchange rates of individuals of the very smallest and very largest organisms are difficult to measure directly. In most instances, populations of microalgae in samples of 1 cm^3 to 1 dm^3 volume, or subsamples of 1–100 cm^2 area cut from macroalgae, provide the material for physiological and biochemical studies. Information on individual microalgal cells are not common, although such measurements are possible (see Chapter 2). This section considers some of the subsampling problems associated with measuring net photosynthesis and respiration rates of macroalgae.

It is common to assess photosynthesis rates of macroalgae by measuring gas exchange rates of "representative" discs removed from the fronds. Sample size is often dictated by the size of the enclosing container. Littler (1979) has recommended the optimum ratio of sample weight to container volume (see above). Considerable variation in photosynthetic and respiratory gas exchange rates of tissue discs is often found. For example, rates of photosynthesis of *Macrocystis pyrifera* measured by different investigators varied from 0.06 to 3.4 mg C·g^{-1} dry wt·h^{-1} and 0.002 to 0.216 g C·m^{-2}·h^{-1} (Arnold and Manley, 1985). Much of this variability appears to be due to the type of tissue under investigation, its physiologic state (i.e., light and nutrient history), physical injury, wounding, sloughing, and the presence or absence of epiphytes (Arnold and Manley, 1985). To overcome the large variability observed in gas exchange rates of different subsamples, Dor and Levy (1987) recommend making sequential observations of photosynthesis and respiration via O$_2$ exchange on the same sample. Although this allows more precise estimates of relative rates of respiration and photosynthesis, it neglects the basic problem of within-plant and between-plant variability.

One way to overcome this subsampling problem is to enclose entire plants or fronds. Hatcher (1977) described a procedure for in situ incubation of macroalgae on the seafloor. Algae were placed in a 76 dm^3 PVC container (30 × 100 cm). Provision was made for filtering seawater to exclude planktonic organisms, continuous mixing to minimize diffusive boundary layers at the plant surface, replacement of seawater during 24-hr incubations to avoid extreme changes in water chemistry, and reduction of O_2 by bubbling with N_2 gas to avoid supersaturation. Enclosures have also been used in experiments with freshwater algae. For example, Kairesalo et al. (1987) used 56 or 112 dm^3 (40 × 40 × 35 or 70 cm) acrylic plastic chambers to enclose intact *Nitella opaca* (Charophyceae).

Gerard (1986) measured net changes in O_2 concentration during 1–2 hr in situ incubations of a single blade or apex of *Macrocystis pyrifera* enclosed in 12–17 dm^3 of ambient seawater in a 21-dm^3 capacity polyethylene bag. In addition to overcoming problems associated with subsampling (i.e., obtaining representative samples and avoiding potential metabolic shocks in damaged tissues), enclosing a sample in a plastic bag can maintain a more realistic turbulence regime (Gust, 1977), which may be important in controlling diffusive fluxes of dissolved gases and nutrients. Losses of O_2 through plastic may, however, introduce an error into measurements of photosynthesis and respiration, although these were found to be negligible provided that the gradient across the bag remained < 50 mmol·m^{-3} (Gerard, 1986).

It is not always possible or desirable to enclose entire plants in chambers for gas exchange measurements, and it may be necessary to employ excised tissue for some rate measurements. Two factors must be considered in using tissue discs: the effects of wounding on gas exchange characteristics and intrablade variability (Arnold and Manley, 1985). Wounding may increase rates of extracellular product formation and oxygen consumption. For example, rates of O_2 consumption were five times higher in freshly cut material than in intact *Laminaria* sp. (Hatcher, 1977), 3.4 times higher in freshly cut than in intact *Macrocystis pyrifera* (Arnold and Manley, 1985) and 1.4 times higher in blade discs of *Laminaria saccharina* compared to intact plants (Gerard, 1988). Release of phenolics from damaged tissue has been suggested as a possible reason for enhanced O_2 consumption. Bidwell and McLachlan (1985) using "moist air suspensions" report that an O_2-dependent respiration that could be confused with photorespiration was actually an artifact due to wounding. Wounding appears to have less of an effect on net photosynthesis, with increases of < 10% reported for *Macrocystis pyrifera* immediately following tissue excision (Arnold and Manley, 1985) and decreases of approximately 25% in blade discs of *Laminaria saccharina* (Gerard, 1988). Prolonged incubation of tissue discs generally leads to rates of photosynthesis and respiration that are lower than those observed in freshly excised material (Hatcher, 1977; Arnold and Manley, 1985).

Within-blade variability of photosynthesis and respiration rates have been reported for *Laminaria* spp. (Hatcher, 1977; Kuppers and Kremer, 1978) and *Macrocystis* spp. (Arnold and Manley, 1985). Differences in

Table 1.5.   Rates of photosynthesis and respiration in *Macrocystis pyrifera*.

	Photosynthesis		Respiration	
	$(mmol\ O_2 \cdot g^{-1}\ dry\ wt \cdot h^{-1})$			
	Mean	SD	Mean	SD
Apical scimitar	0.32	±0.11	0.10	±0.033
Blade	0.26	±0.04	0.056	±0.021
Sporophylls	0.12	±0.05	0.036	±0.018
Stipe	0.013	±0.005	0.011	±0.005
Holdfast	>0.000		0.015	±0.003

Adapted from Arnold and Manley (1985).

rates of photosynthesis and respiration are perhaps best documented for the morphologically complex *Macrocystis pyrifera* (Table 1.5). The highest rates were found in apical scimitar and blades, which make up the bulk of the photosynthetic tissue, whereas lower rates were found in sporophylls, stipe, and holdfast (Arnold and Manley, 1985). Within- and between-blade variability in *Macrocystis pyrifera* was high, two- to fivefold differences in mass-specific photosynthesis rates along blades and up to fivefold differences between blades from mature plants (Arnold and Manley, 1985). It is perhaps not surprising that gas exchange rates vary widely between different tissues, given the structural and morphological complexity of large seaweeds.

Comparison of O_2 and CO_2 Exchange

Many investigations of the photosynthetic or respiratory metabolism of algae employ either O_2 exchange or TCO_2 exchange but rarely both in combination, even though it has long been recognized that O_2 and TCO_2 fluxes may not be equivalent. The ratio of the rates of oxygen evolution to carbon dioxide assimilation during photosynthesis is defined as the photosynthetic quotient (designated PQ). Similarly, a respiratory quotient (designated RQ) can be defined as the ratio of oxygen consumption to carbon dioxide evolution during respiration. Photosynthetic and respiratory quotients of unity are commonly assumed when direct observations are lacking, based on the synthesis or remineralization of carbohydrate:

$$CO_2 + H_2O \leftrightarrow O_2 + CH_2O \qquad (1.20)$$

Although a photosynthetic quotient of unity is often assumed, there are many situations in which large deviations (± 50%) are to be expected.

Oxygen and inorganic carbon exchanges are the result of a number of biochemical reactions, some of which operate largely independently. Oxygen evolution provides a measure of the electrons available for biological redox reactions. In light, the evolution of an O_2 molecule is associated with the flow of four electrons from H_2O to reduce $2NADP^+$ to $2NADPH$ during linear electron flow. Coupled to this electron flow is a variable amount of ATP synthesis of < 3 $ATP/O_2 < 3-4$ ATP. Inorganic carbon is only one of the electron acceptors that compete for the reductant produced as a consequence of linear electron flow. Other electron acceptors include nitrate, nitrite, sulphate, or molecular oxygen. Nitrite reduction to ammonium can account for approximately 40% of photoreduction at low irradiances, decreasing to approximately 10% at irradiances that saturate photosynthetic oxygen evolution (Curtis and Megard, 1987). Changes in the photosynthetic quotient with irradiance may largely reflect a changing stoichiometry of nitrate and/or nitrite reduction to CO_2 fixation (Megard et al., 1985).

Photosynthetic quotients exceeding unity are expected for growth of micro- and macroalgae based on stoichiometries of element exchanges. Reduction of nitrate to ammonium, which requires the transfer of eight electrons for each nitrogen atom reduced, accounts in part for photosynthetic quotients that exceed unity in algae. These additional electrons can be accounted for in a mass balance equation (Falkowski et al., 1985):

$$nCO_2 + (n + 1) H_2O + HNO_3 \Rightarrow (CH_2O)_n(NH_3) + (n + 2)O_2 \quad (1.21)$$

Given the assumptions that nitrate is the inorganic nitrogen source and carbon is reduced to the level of carbohydrate, mass balance equations for microalgal and macroalgal growth (Atkinson and Smith, 1983) are as follows.

For microalgae:

$$106\ CO_2 + 122\ H_2O + 16\ HNO_3 + H_3PO_4 \rightarrow \quad (1.22)$$
$$(CH_2O)_{106}(NH_3)_{16}(H_3PO_4) + 138\ O_2$$

For macroalgae:

$$550\ CO_2 + 580\ H_2O + 30\ HNO_3 + H_3PO_4 \rightarrow \quad (1.23)$$
$$(CH_2O)_{550}(NH_3)_{30}(H_3PO_4) + 610\ O_2$$

Although considered typical of micro- and macroalgal growth, the preceding equations do not apply in all situations. The Redfield stoichiometry (Eq. 1.22) was derived for the average elemental composition of organic matter in the oceans, and significant local variations above and below these proportions are found. The C:N and C:P ratios in phytoplankton can vary widely from the Redfield proportions depending on nutrient availability and irradiance with molar C:N ratios ranging 4–25, and C:P ratios ranging 75–625 (Goldman, 1980). The data for macroalgae span a range of C:N ratios of < 10 to 60 and C:P ratios of 140–3550 (Atkinson and Smith, 1983). Both the irradiance and availability of inorganic nutrients affect elemental composition ratios. Low C:N and C:P ratios are found under nutrient-sufficient, but light-limited conditions, whereas high C:N and C:P ratios are indicative of light-saturated, but nitrogen-limited or phosphorus-limited conditions. The N:P ratio reflects the relative availability of these two important inorganic nutrients and may vary from 5–115 in microalgae (Goldman et al., 1979) or 5–180 in macroalgae (Atkinson and Smith, 1983).

Further increases in the PQ over those calculated on the basis of nitrate reduction (Eq. 1.21) can be attributed to the fact that organic carbon in lipids and proteins is more highly reduced than organic carbon in carbohydrate. Protein accounts for 10% to 50% of microalgal dry weight, and lipid accounts for 15% to 70% of dry weight (Hitchcock, 1982; Shifrin and Chisholm, 1981; Ditullio and Laws, 1986). In contrast, macroalgae are comprised predominantly of carbohydrate (80%), with protein accounting for only 15% and lipid < 5% of ash-free dry weight (Atkinson and Smith, 1983).

Calculations based on the biochemical composition of microalgae lead to estimates of the photosynthetic quotient for net cell synthesis of 1.1–1.2 when a reduced nitrogen source such as urea or ammonia is available and 1.4 when nitrate is the nitrogen source (Cramer and Myers, 1948; Myers, 1980; Pirt, 1986). Nitrogen-fixing cells should have a PQ of approximately 1.24 (Pirt, 1986). It is thus somewhat paradoxical that early observations of the PQ obtained from gas exchange measurements often yielded values close to unity, consistent with carbohydrate as the major product of photosynthesis (Ryther, 1956). The low PQs observed by early investigators are not representative of balanced algal growth, and Myers and Cramer (1948) considered a PQ of unity to be a consequence of measuring photosynthesis at very high irradiances in cultures that were acclimated to low light. These are the conditions under which a large amount of carbohydrate synthesis and relatively small amounts of protein and lipid synthesis are expected. Exponentially growing phytoplankton cultures have PQs consistent with calculations based on the elemental composition and nitrogen source (Table 1.6). However, large variations may be observed in stationary-phase cultures or immediately following subculturing (Thomas, 1964).

Photosynthetic quotients of 1.21 ± 0.06 (Buesa, 1980) and 1.17 ± 0.02

Table 1.6.   Stoichiometries of phytoplankton growth on different nitrogen sources.

$0.89\ CO_2 + 0.61\ H_2O + 0.127\ NH_3$ $\Rightarrow 0.89$ C-mol biomass + O_2	PQ $= 1.12$
	C:N $= 7.01$
$0.82\ CO_2 + 0.62\ H_2O + 0.067\ N_2H_4CO \Rightarrow 0.89$ C-mol biomass + O_2	PQ $= 1.12$
	C:N $= 6.64$
$0.71\ CO_2 + 0.59\ H_2O + 0.101\ HNO_3$ $\Rightarrow 0.71$ C-mol biomass + O_2	PQ $= 1.41$
	C:N $= 7.02$
$0.81\ CO_2 + 0.73\ H_2O + 0.058\ N_2$ $\Rightarrow 0.81$ C-mol biomass + O_2	PQ $= 1.23$
	C:N $= 6.98$

Data from Myers (1980) and Pirt (1986).

(Axelsson, 1988) have been obtained for macroalgae from the Chlorophyceae, Phaeophyceae, and Rhodophyceae, with no significant difference in the values for different taxonomic groups. These values were obtained at air-equilibrium O_2 concentrations in natural seawater. Buesa (1980) measured O_2 and CO_2 exchange using a Warburg manometer and Axelsson (1988) measured pH and O_2 in a well-mixed flow through chamber. In both cases, diffusive boundary layers at the algal surface are expected to be minimized. Similar values have been obtained by other investigators (Oates and Murray, 1983).

Variations in PQ are not always explicable in terms of nitrogen metabolism, and other metabolic processes may also have to be considered. Photorespiratory metabolism can contribute to PQs significantly less than unity. Littler (1979) found that the PQ decreased from unity at low O_2 ($< 1\ \mu mol \cdot mol^{-1}$) to 0.5 at air-saturated O_2 ($6.6\ \mu mol \cdot mol^{-1}$) and to 0.25 at supersaturated O_2 ($15.5\ \mu mol \cdot mol^{-1}$) in *Ulva lobata* at 20°C and an irradiance of $140\ \mu mol \cdot m^{-2} \cdot s^{-1}$. Burris (1981) demonstrated an inverse relationship between PQ and the partial pressure of O_2 in three marine microalgae at a temperature of 20°C and an irradiance of 1.3×10^{17} photons $\cdot cm^{-2} \cdot s^{-1}$. In *Glenodinium* sp. and *Pavlova lutheri*, PQ decreased linearly from 1.2–1.8 at pO_2 of 0.1 atm ($110\ \mu mol\ O_2 \cdot dm^{-3}$) to 0.5–0.7 at pO_2 of 0.3 atm ($330\ \mu mol\ O_2 \cdot dm^{-3}$). Kaplan and Berry (1981) found that decreases in the PQ of *Chlamydomonas reinhardtii* were correlated with increases in the excretion of glycolate, which is one of the products of ribulose bisphosphate carboxylase-oxygenase (RUBISCO) oxygenase activity. In contrast, Bidwell and McLachlan (1985) found no evidence for photorespiration except in an atmosphere with simultaneously elevated O_2 and very low CO_2 ($50 \times 10^{-4}\ cm^3 \cdot dm^{-3}$), although an O_2-sensitive wound respiration that could potentially be confused with photorespiration was observed. Variations in the PQ of aquatic macrophytes (with PQ often less than unity) may result from differences in the accumulation of oxygen in the lacunal gas-space (Pokorny, et al., 1989).

Changes in Biomass

The autotrophic biomass or standing stock is the concentration of plant material per unit volume $(g \cdot m^{-3})$ or per unit area $(g \cdot m^{-2})$. Biomass is usually measured as wet weight, dry weight, ash-free dry weight, or organic carbon, although a number of other variables can also be employed. In studies of phytoplankton photosynthesis, chlorophyll concentration is the most readily and unambiguously measured biomass indicator. The increase in autotrophic biomass over a given time interval is the yield, which, like standing stock, has units of $g \cdot m^{-2}$ or $g \cdot m^{-3}$. Under natural conditions the yield is determined by the difference between net algal photosynthesis and losses to grazers, export, and excretion, where all variables are reported in the same units.

$$\text{Yield} = \text{Net Photosynthesis} - \text{Excretion} - \text{Grazing} - \text{Export} \quad (1.24)$$

Problems in balancing Eq. 1.24 can arise if measurements of the different variables are obtained in different units. For example, net photosynthesis is often obtained from oxygen evolution, excretion from release of dissolved organic ^{14}C, and yield from the accumulation of dry weight. A parameter related to the yield is the net production rate (Strickland, 1960), defined as the rate of increase of autotrophic biomass. Yield can be considered as the integrated value of net production rate over time. In axenic cultures, the export and grazing terms are by definition zero, and the increase in biomass provides an independent check on measurements of net gas exchange and excretion.

In field samples, export can be eliminated by enclosing a sample, but the problem of eliminating grazing remains. Using enclosures may also introduce other artifacts such as a reduction in turbulence, adsorption of nutrients on container surfaces with consequent changes in the chemical composition of the aqueous phase, and elimination of external nutrient sources. With macroalgae, grazing can be reduced by physically removing large macroscopic grazers or by choosing samples that are free from grazers. It is difficult to separate microalgae from protozoan grazers, but a protocol allowing phytoplankton growth to be uncoupled from microscopic grazers has recently been developed (Landry and Hassett, 1982).

Coupling of Growth and Photosynthesis

Changes in biomass are also used to calculate the specific growth rate:

$$\mu = \ln \left[B_f / B_0 \right] / \delta t = \ln \left[(B_0 + Y) / B_0 \right] / \delta t \quad (1.25)$$

where B_0 is the initial biomass, B_f is the final biomass, $Y = B_f - B_0$ is the yield, and δt is the time interval between measurements of B_0 and B_f. Growth is said to be balanced when the specific growth rates (Eq. 1.25) obtained from all measured biomass variables are equal. Differences in the specific rates of synthesis of various biomass variables typifies the unbalanced growth that occurs following changes in environmental conditions, such as a shift in irradiance or exhaustion of nutrients.

Algal growth is not always directly coupled to the rate of photosynthesis. Uncoupling of growth and photosynthesis is evident on diurnal time scales, with net photosynthesis and carbohydrate accumulation limited to the illuminated part of the day, whereas net protein synthesis often continues in darkness (Cuhel et al., 1984). Uncoupling growth and photosynthesis is also evident when nutrient-limited algae are exposed to elevated nutrient concentrations. Storage carbohydrates and lipids are accumulated under energy-sufficient but nutrient-limited conditions and can be rapidly mobilized following nutrient additions. Ammonium assimilation into low-molecular-weight compounds occurs at a faster rate than macromolecular synthesis when nitrogen-limited algae are provided with increased ammonium concentrations. This occurs because the maximum cell-specific ammonium uptake rate is largely independent of growth rate, whereas the rate of macromolecular synthesis is nearly proportional to growth rate in ammonium-limited chemostat cultures (Zehr et al., 1988; Kanda and Hattori, 1988). As a consequence, the pool of low-molecular-weight nitrogen compounds shows a transient increase following ammonium addition to nitrogen-deficient cells, and the ratio of short-term ammonium uptake to nitrogen incorporation into macromolecules provides a good indicator of nitrogen deficiency. This increase can be measured by adding $^{15}NH_4$ to a sample and following the accumulation of ^{15}N label in low-molecular-weight (hot ethanol extract) and macromolecular (hot ethanol insoluble) pools (Kanda and Hattori, 1988). Using this approach, Kanda et al. (1988) found both nitrogen-sufficient and nitrogen-deficient phytoplankton populations in oligotrophic waters of the western North Pacific near Japan.

An extreme case of the uncoupling of photosynthesis and growth has been observed in *Laminaria solidungula* from the high Arctic (Chapman and Lindley, 1980; Dunton, 1985). Carbohydrate reserves (laminarin) are built up during the summer when light is available but inorganic nutrients are low, and growth occurs at the expense of stored carbohydrates over winter when nutrients are available but light levels are low. Less extreme cases of uncoupling of photosynthesis, nutrient assimilation and growth have been documented for other macroalgae. The most common pattern is similar to that observed in *L. solidungulu* and involves an accumulation of carbohydrate reserves during the summer when temperatures are high and nutrients are low and an accumulation of tissue nitrogen during the winter when growth rate is limited by temperature but nutrients are available (Duke et al., 1987).

Community Metabolism

All of the methods that require enclosing the sample also at least partially isolate the enclosed organisms from the rest of their environment. This isolation may have several effects including (1) reducing vertical and horizontal mixing, (2) reducing turbulent mixing, particularly in studies of macrophytes and benthic microalgae, (3) removing large grazers and sources of regenerated nutrients, and (4) introducing contaminants that may alter gas exchange rates. These problems can be overcome if gas exchange rates can be estimated in unconfined conditions.

Photosynthetic and respiratory activity changes oxygen concentration and pH within a waterbody. Oxygen accumulates, TCO_2 is consumed, and pH rises when photosynthesis exceeds respiration, whereas O_2 is consumed, TCO_2 is released, and pH drops when respiration exceeds photosynthesis. These processes produce patterns of in situ O_2, TCO_2, and pH variations on time scales ranging from hours to weeks that can often be used to calculate biological gas exchange rates and net productivity of aquatic ecosystems (Kinsey, 1985). However, these patterns are strongly influenced by atmospheric exchange, which tends to reduce in situ variations in shallow, well-mixed waterbodies (Hartman and Hammond, 1985), whereas stratification in deeper water columns is reduced by vertical mixing.

The method for estimating community metabolic rates from diurnal changes in O_2 and/or TCO_2 in freshwater and marine environments is often associated with the work of Odum and coworkers (Odum and Odum, 1955; Odum, 1956; Odum and Hoskins, 1958; Park et al., 1958; Beyers and Odum, 1959). Although Odum is often credited with developing the method, the approach was first used to estimate phytoplankton productivity by workers in the Soviet Union (Brujewicz, 1936; Vinberg and Yarovitzina, 1939; Vinberg, 1940). Early investigations were also conducted on Wisconsin lakes by Juday et al. (1943) and in African waters by Talling (1957a, b). A review of the early work can be found in Talling (1957b). A summary of studies conducted between 1958 and 1980 can be found in Johnson et al. (1981) and references to more recent work can be found in Kinsey (1985). With increases in the accuracy of O_2 and TCO_2 determinations (Table 1.1), open-water techniques are being used more often in studies of open-ocean primary production (Tijssen, 1979; Oudot, 1989).

Open-water methods for estimating primary production have been advocated by some investigators, because this approach eliminates any potential artifacts associated with enclosing planktonic or benthic organisms within bottles or bell jars. The deficiencies inherent in open systems include a lower level of precision, which is associated with heterogeneity in the water mass under investigation, a requirement to estimate the fluxes associated with physical exchange across the air–water interface, and the need to measure exchanges with other water masses. The spatial and temporal resolution of a sampling program must accommodate the spatial and temporal variability of O_2 or TCO_2 concentrations in the environment examined if accurate rates of community photosynthesis and respiration are to be obtained. In general, constraints on manpower and equipment limit the spatial resolution to one or a few locations in the horizontal, one or a few depths in the vertical, and the temporal resolution to several hours between samples. However, automation of data collection and computer-aided analysis can greatly extend the application of open-water techniques (Griffith et al., 1987).

Homogeneity in both the physical environment and in the biological community is required if open-water techniques are to be applied successfully. In addition, biological oxygen-producing and -consuming processes must exceed rates of physical exchange across the system boundaries if meaningful data are to be obtained. Otherwise, where physical processes dominate the O_2 budget, small errors in estimates of physical exchange can lead to very large errors in calculated biological fluxes (Kemp and Boynton, 1980).

The O_2 budget for a well-mixed, horizontally homogeneous water column can be expressed as

$$d[O_2] / dt = P - R + k ([O_2] - [O_2]_{eq}) / D \qquad (1.26)$$

where $[O_2]$ is the dissolved oxygen concentration ($mmol \cdot dm^{-3}$), P is the rate of photosynthetic O_2 production ($mmol \cdot dm^{-3} \cdot s^{-1}$), R is the rate of respiratory O_2 consumption ($mmol \cdot dm^{-3} \cdot s^{-1}$), $[O_2]_{eq}$ is the equilibrium dissolved oxygen concentration predicted from temperature and salinity, k is the atmospheric exchange coefficient ($m \cdot s^{-1}$), and D is the depth of the mixed layer (m). The exchange coefficient k is strongly dependent on wind speed (V) with $k \approx V^2$ (Broeker and Peng, 1982). The importance of atmospheric exchange is greatest in shallow mixed layers and when $[O_2]$ deviates greatly from $[O_2]_{eq}$. Hartman and Hammond (1985) recently reexamined the relationship between wind speed and the gas exchange coefficient and proposed the following empirical relationship, which fits available data to within 20% for wind speeds of 3–12 $m \cdot s^{-1}$:

$$K = 34.6 \text{ (s cm}^{-2}) R_{nu}(D_{m20})^{0.5}(U_{10})^{1.5} \tag{1.27}$$

where K is the liquid-phase gas-transfer coefficient (m·d^{-1}), R_{nu} is the ratio of the kinematic viscosity of pure water at 20°C to the kinematic viscosity of water at the measured salinity and temperature (dimensionless), D_{m20} is the molecular diffusivity of the gas of interest at 20°C (cm^2·s^{-1}) and U_{10} is the wind speed 10 m above the water surface (m·s^{-1}). Typical values for R_{nu} and D_{m20} are given in Hartman and Hammon (1985).

Equation 1.26 cannot be used for stratified water columns in which the depth dependence of [O$_2$], P, and R must be explicitly recognized (Talling, 1957). In regions isolated from the air–water interface in which there is little advective transport of water, the O$_2$ budget can be expressed as

$$d[O_2]_z / dt = P_z - R_z - K_z d^2[O_2]_z / dz^2 \tag{1.28}$$

where the depth dependence is made explicit by the subscript z and the physical exchange is assumed to be parameterized by a diffusion term with an eddy diffusion coefficient K_z. The vertical profile of O$_2$ often shows a subsurface maximum in the pycnocline in open ocean and lake waters. This subsurface oxygen maximum is normally located above the chlorophyll maximum layer in the open sea and changes of [O$_2$] within this layer can be used to infer rates of net photosynthesis (Shulenberger and Reid, 1981).

Open-water measurements have also been made in flowing waters. In this case, gas exchange rates can be calculated from

$$\text{Areal Production} = \Delta O_2 V_d / l \tag{1.29}$$

where areal production has units of moles of O$_2$ per square meter per second, ΔO_2 is the difference in concentration between the two points (in mol O$_2$·m^{-3}), V is the velocity (in m^3·s^{-1}), d is the water depth (in m) and l is the transect length (in m). This equation applies to closed channels in which the residence time can be estimated from the mean fluid velocity and the transect length.

A major advantage of measuring open-water TCO$_2$ fluxes rather than O$_2$ fluxes is that TCO$_2$ exchanges slowly with the atmosphere in most natural waters and thus atmospheric corrections are minimal or unnecessary. An additional advantage in studies on coral reefs is that calcification and organic carbon fluxes can be obtained (Kinsey, 1985). However, the TCO$_2$ flux includes contributions from chemosynthetic CO$_2$ fixation and fermentation in addition to oxygenic photosynthesis and respiration. A disadvantage of measuring TCO$_2$ fluxes is that the analytical precision is generally less than can be obtained for O_2 measurements. Although high precision can be obtained with the coulometric titration (Johnson et al., 1987), most estimates of TCO_2 fluxes have been obtained indirectly through measurement of pH changes.

Comparison of O_2 and TCO_2 fluxes in open-water systems leads to PQs and RQs ranging from 0 to 5 (Park et al., 1958; Beyers, 1963; Johnson et al., 1981; Oviatt et al., 1986), although the central tendency is usually approximately 1–1.5 (Oviatt et al., 1986). A number of processes may account for this wide range. As already mentioned, the effects of nitrate assimilation and photorespiration by aquatic plants can account for a range of PQs from 0 to 2. Additional processes include denitrification, in which NO_3^- acts as the electron acceptor for respiratory pathways in place of O_2, and calcium carbonate precipitation, which leads to a reduction of TCO_2 without affecting O_2. In addition, environmental variations in pH, light, or temperature may contribute to the wide range of measured values (Pokorny et al., 1989).

Open-water measurements need not be limited to a diel time scale. Longer term changes in O_2, TCO_2, and inorganic nutrient (NO_3^-, PO_4^{3-}) concentrations can be used to estimate net seasonal and longer term changes in community metabolism and the rates of exchange of gases between the ocean and the atmosphere (Redfield, 1948; Johnson et al., 1979; Jenkins and Goldman, 1985; Spitzer and Jenkins, 1989). However, a discussion of this approach is beyond the scope of this book and will not be pursued further.

Abiological Oxygen Consumption

In using O_2 for measuring photosynthetic gas exchange, one complicating factor which is particularly evident in coastal or freshwaters with a high humic acid content and may also be a problem in open ocean waters, is photochemical oxygen consumption. The observation of O_2 undersaturation in highly colored freshwaters characterized by low biological activity prompted some investigators to suggest that purely chemical oxidations may be responsible for a portion of the O_2 consumption in natural waters (Table 1.7).

One candidate for this chemical oxidation is a reaction cycle involving photoreduction of Fe(III) to Fe(II) by humic acids and oxidation of Fe(II) to Fe(III) by dissolved oxygen (Stumm and Morgan, 1970):

$$Fe(III) - org \; complex \Rightarrow Fe_2^+ + oxidized \; org \qquad (1.30)$$

$$Fe_2^+ + org + 0.25 \; O_2 \Rightarrow Fe(III) - org \; complex \qquad (1.31)$$

Miles and Brezonik (1981) demonstrated light-enhanced O_2 consumption in highly colored lake waters that had been filtered to remove particulate matter (Whatman GF/A) and poisoned with mercuric chloride to prevent microbial respiration. They found that the rate of O_2 consumption was greater in waters with higher humic acid content, was faster at higher pH, and increased with increasing total Fe concentration. The rates of abiological oxygen consumption in the dark were only 20% of the light-enhanced rates measured at an irradiance of 500 $\mu mol \cdot m^{-2} \cdot s^{-1}$. Oxygen consumption in the light was accompanied by CO_2 evolution with a stoichiometry of $1/2O_2:CO_2$, consistent with Fe-mediated photooxidation of aminopolycarboxylates (Miles and Brezonik, 1981):

$$RHNCH_2COOH + 0.5O_2 \Rightarrow RNH_2 + CO_2 + CH_2O \qquad (1.32)$$

Iron-mediated, light-dependent O_2 consumption has been demonstrated for several organic compounds containing amino, carboxyl, and hydroxyl groups (Stumm and Morgan, 1970; Miles and Brezonik, 1981), whereas some other compounds, such as tannic acid, promote O_2 consumption in both light and darkness in the pH range 2–11 (Miles and Brezonik, 1981).

Production of superoxide ($O_2^-\cdot$) by near ultraviolet light (< 400 nm) and consumption of this $O_2^-\cdot$ in the oxidation of organic matter provides another mechanism for photochemical oxygen consumption (Zepp et al., 1977, Zafiriou et al., 1984). Lanne et al. (1985) concluded that photochemical processes accounted for O_2 consumption in seawater filtered through double layers of Whatman GF/F filters or 0.2-μm pore membrane filters and implicated singlet oxygen production as a factor in the photooxidation. Higher rates of O_2 consumption were observed in quartz bottles and blood transfusion bags than in glass bottles, presumably because of greater penetration of ultraviolet radiation in the former. Rates of photochemical O_2 consumption varied from about 0.2 $\mu mol \cdot dm^{-3} \cdot h^{-1}$ in the Sargasso Sea to 165 $\mu mol \cdot dm^3 \cdot h^{-1}$ in the eutrophic Lake Tjeuker-

Table 1.7. Comparison of photosynthetic oxygen production and photochemical oxygen consumption rates in selected natural waters.

Aquatic Ecosystem	Net Photosynthetic O_2 Production (mmol $O_2 \cdot m^{-2} \cdot d^{-1}$)	Photochemical O_2 Consumption (mmol $O_2 \cdot m^{-2} \cdot d^{-1}$)
Caribbean Sea	5	1
North Sea	10	4
Ems-Dollart Estuary	4	0.2
Lake Tjeukemeer	19	3

Adapted from Lanne et al. (1985).

meer in the Netherlands (Table 1.7) and accounted for the consumption of up to 40% of net photosynthetic O_2 evolution (Lanne et al., 1985).

Diurnal trends in a number of photochemical products have been demonstrated in natural seawater: these include aldehydes and ketones (Mopper and Stahovec, 1986), hydrogen peroxide (Zika et al., 1985), carbon monoxide (Conrad and Seiler, 1980), and carbonyl sulfide (Ferek and Andreae, 1984). Kieber and Mopper (1987) recently demonstrated photochemical formation of glyoxylic and pyruvic acid at rates comparable to biological production and consumption rates in ocean waters ranging from the oligotrophic Sargasso Sea to Florida Bay. In contrast to conventional wisdom, which assumes that virtually all of the cycling of low-molecular-weight compounds in the sea is mediated by biological processes, recent studies suggest an important role for photochemical reactions in the natural cycles of these compounds (Zafiriou et al., 1984; Kieber and Mopper, 1987).

Chapter 2

Using Isotopes to Measure Gas Exchange

Tracers have long been used in photosynthesis experiments to provide information that cannot be obtained by net gas exchange techniques. The radioactive tracer ^{14}C, was instrumental in determining the biochemical pathways of CO_2 fixation (Bassham and Calvin, 1957). In addition, assimilation of inorganic ^{14}C has become a standard technique for measuring photosynthesis of unicellular aquatic plants, largely replacing net oxygen exchange, particularly under oligotrophic conditions. The stable carbon isotope ^{13}C can substitute for ^{14}C under some conditions and measurements of natural abundance of ^{13}C relative to [12]C (the carbon-isotope ratio $\delta^{13}C$) have provided considerable insight into photosynthetic mechanisms and form(s) of inorganic carbon assimilated by plants. Measurements of $\delta^{13}C$ have also been used to elucidate the physical and biochemical rate-limiting steps in photosynthetic carbon fixation. The stable oxygen isotope [18]O has been useful in evaluating the roles of respiration, photorespiration, and the Mehler reaction in illuminated plants. Tritium (^{3}H) has not proven to be a useful radioisotope for examining gas exchange, but tritiated amino acids and sugars have been used to examine algal osmotrophy (Rivkin and Putt, 1987) and [3]H_2O has been used in experiments to estimate rates of protein turnover (Richards and Thurston, 1980).

[14]C–CO_2 Assimilation

The sensitivity of net gas exchange techniques for measuring inorganic carbon assimilation is ultimately limited by the precision of the TCO_2 analysis. The most demanding environment for measuring net carbon exchanges is the oligotrophic open ocean in which TCO_2 concentration is ≈ 2.4 mol·m^{-3}, but photosynthesis rates rarely exceed 0.5 mmol C·m^{-3}·d^{-1} (Hayward et al., 1983). Thus, a TCO_2 method capable of resolving differences of approximately 0.02% is needed to measure net daily TCO_2 exchange. Even the most sensitive coulometric technique (Johnson et al., 1985) does not have the sensitivity required to measure these TCO_2 changes. In contrast, the ^{14}C technique can reliably measure photosynthesis rates even in the most oligotrophic ocean regions provided sufficient care is taken to insure the samples are not contaminated by trace metals or other inhibitory substances. One indication of the sensitivity of the ^{14}C technique is that photosynthesis rates of individual phytoplankton cells can be measured (Rivkin and Seliger, 1981; Rivkin, 1985).

Steemann-Nielsen (1952) introduced the ^{14}C technique to phytoplankton studies to overcome the sensitivity problem encountered with conventional gas exchange measurements. The basic method is relatively simple. A known amount of inorganic ^{14}C-bicarbonate is introduced in trace amounts into a bottle containing a water sample. The sample is incubated for a specific period of time, which usually ranges between 1 hr and 1 d. At the end of the incubation, the inorganic ^{14}C is separated from the organic ^{14}C. A standard method for this separation involves acidifying a sample to pH 2 and allowing the inorganic $^{14}CO_2$ to escape (Schindler et al., 1972). Acidification and purging of inorganic ^{14}C should be carried out in the vial that will be used for scintillation counting to avoid loss of organic ^{14}C (Theodórsson and Bjarnason, 1975). The carbon assimilation rate is calculated from the measured organic ^{14}C activity, the total activity of inorganic ^{14}C added to the sample, and the total inorganic carbon concentration in the sample:

$$(\delta C / \delta t) = f (dpm_{org} / dpm_{tot}) [TCO_2] (1/\delta t) \qquad (2.1)$$

where $\delta C/\delta t$ is the carbon assimilation rate, f is an isotope discrimination factor generally assumed to equal 1.06 (dimensionless), dpm_{org} is the ^{14}C activity in organic matter (in disintegrations per minute), dpm_{tot} is the total inorganic ^{14}C activity added to the sample, $[TCO_2]$ is the total inorganic carbon concentration, and δt is the duration of the incubation.

Although ^{14}C activities are often reported as dpm, the S.I. unit for radioactivity is the Becquerel (Bq) which equals one disintegration per second. Many users and suppliers of radioactive isotopes retain the use of the Curie (Ci) in specifying the activity of a radioactive element. A Curie (Ci) is defined as the amount of radioactive material in which the number of disintegrations per minute is the same as that of 1 g of radium. The relationship between these different units of radioactivity are 1 Ci = 3.7 10^{10} Bq = 2.22 10^{12} dpm. Finally, the raw data obtained from a scintillation counter is often reported in units of counts per minutes (cpm). The number of counts per minute that a scintillation counter detects is related to the activity expressed in disintegrations per minute through the counting efficiency (dpm = cpm/counting efficiency).

The sensitivity of the ^{14}C method is limited by the accuracy of the measurements of ^{14}C activity and total dissolved inorganic carbon concentration. In the original technique, Geiger Muller counting was used to measure ^{14}C activity and a number of corrections had to be made for sample geometry and quenching (Vollenweider, 1969), but these corrections were subject to some uncertainty, limiting the accuracy of the method. Liquid scintillation counting was introduced by Schindler (1966) to replace Geiger-Muller counting for determining ^{14}C activity resulting in an increase in the accuracy and versatility of the method over Steemann-Nielsen's original protocol.

Some modifications to experimental protocol are required for samples of macroalgae (Arnold and Littler, 1985). Weighed subsamples are usually taken from a thallus or frond that has been incubated in the presence of ^{14}C, and the plant material must be solubilized prior to liquid scintillation counting (Beer et al., 1982; Lewis et al., 1982; Fillrin and Hough, 1984). Berger and Bate (1986) report that tissue solubilizers may not effectively extract ^{14}C from structural material and recommend dry combustion to obtain complete recovery of organic ^{14}C. They found that only 10% of organic ^{14}C was extracted in dimethyl sulfoxide (DMSO) and 30% in Instagel® or Filtercount®, but that recovery was virtually 100% when samples were combusted. Although collecting subsamples allows the investigation of physiological variation in photosynthetic metabolism by different parts of the plant, complications arise in attempting to calculate whole-plant productivity from the weights and specific uptake rates of component parts.

Dissolved Organic ^{14}C Production

Dissolved organic ^{14}C (DO^{14}C) production is operationally defined as the acid-stable ^{14}C activity that is found in the filtrate from a fine pore size filter; nominally a 0.45-μm Millipore® cellulose acetate membrane or 0.2-μm Nuclepore® polycarbonate filter is employed. A number of technical problems arise in measuring the small amounts of DO^{14}C produced against a background of potential contaminants, including acid-stable activity in the stock ^{14}C-bicarbonate solution, organic ^{14}C leached from cells disrupted during filtration, and autotrophic and heterotrophic cells or cell fragments that are smaller than the size of the filter pores (Sharp, 1977). Care must also be taken to insure that the sample is completely purged of inorganic ^{14}C (Sharp, 1977). Finally, assimilation of DO^{14}C by heterotrophic bacteria during the course of a ^{14}C uptake experiment may be substantial in natural samples (Laanbroek et al., 1985; Fuhrman, 1987; Sundh, 1989), which would lead to an underestimate of the amount of DO^{14}C produced.

Filtration must be carried out with a low vacuum to avoid rupturing fragile cells, and membrane filters or polycarbonate sieves may cause less damage than glass fiber filters (Goldman and Dennett, 1985). For microalgal samples, the pore size of the filter must be small enough to retain all photosynthetic cells, and several comparisons of the efficiency of various types of filters for retaining chlorophyll a, particulate ^{14}C, and particulate organic carbon have been made (Sheldon, 1972; Salonen, 1979; Li, 1986; Taguchi and Laws, 1988). Recent investigations have demonstrated the importance in both fresh and ocean waters of chloro-phyll-a-containing picoplankton (Stockner, 1988), which include cyano-bacteria, chloroxybacteria, and eucaryotes. This size class ranges from 0.2 to 2 μm in diameter and is most often operationally defined as those organisms that pass through a 1.0-μm pore size sieve but are retained by a 0.2 μm pore size sieve or 0.45-μm pore size membrane filter (Li, 1986). Even a 0.2 μm filter does not always retain all chlorophyll-a-containing organisms (Taguchi and Laws, 1988).

Extracellular release of DO^{14}C accounts for a small proportion of photosynthesis in many micro- and macroalgae and appears to be a normal consequence of algal metabolism. Dissolved organic ^{14}C production has been observed to continue at a constant rate during light to dark transitions (Mague et al., 1980), suggesting that this loss of cell material is by passive diffusion. The flux of dissolved organic matter via passive permeation through cell membranes is proportional to the membrane permeability times the concentration gradient across the cell membrane (Bjornsen, 1988). In addition to release via diffusive processes, DO^{14}C can be released to the dissolved pool via active excretion, autolysis, and as a product of grazing activity (Fuhrman, 1987).

Brylinsky (1977) reported release of only 1 to 4% of total organic ^{14}C assimilation as DO^{14}C in experiments with five benthic and one pelagic species of marine macrophytes, including two species of red algae and two species of brown algae. Healthy plants with minimal accumulation of epiphytes were incubated for 2 hr in plexiglass chambers containing added ^{14}C-bicarbonate in situ. Similar low rates of dissolved organic carbon (DOC) production have been reported for green (Craigie et al., 1966; Pregnall, 1983), red (Majak et al., 1966), and brown macroalgae (Harlin and Craigie, 1975). It is possible to measure extracellular release of DOC from macroalgae without recourse to ^{14}C tracer experiments because of the large biomass/volume ratios obtained when whole fronds are immersed in small volumes of seawater. Carlson and Carlson (1984) reported that extracellular release rates equalled $1g \cdot m^{-2}$ substratum$\cdot d^{-1}$, equalling 2 to 5% of daily production by fucoid macroalgae. High rates of extracellular release (30 to 35% of organic ^{14}C production) have been reported for tissues wounded during sample processing (Khailov and Burlakova, 1969) and may not be representative of healthy macrophytes. Extracellular production of dissolved organic matter can also be expected to be significant in senescent tissue (Mann et al., 1980).

Pregnall (1983) examined the environmental factors which affect DOC release in *Enteromorpha prolifera*. Dissolved organic ^{14}C release accounted for a small fraction (2 to 6%) of total ^{14}C assimilation in submerged plants at 30‰ salinity, increasing linearly with decreasing salinity to 16% at 5‰. However, a large fraction of previously fixed ^{14}C was released as DOC in a pulse following reimmersion of plants that had been exposed to desiccation. Submergence in freshwater led to fivefold greater losses of DOC than reimmersion in seawater. Although DOC release was a small proportion of total photosynthesis in submerged plants, the amount of DOC released increased dramatically in plants exposed to cycles of desiccation and inundation. As periods of desiccation and inundation are quite common for intertidal species such as *Enteromorpha* this warrants further examination.

Both high- and low-molecular-weight (Table 2.1) compounds contribute to DOC release, indicating that passive permeation is not the only source of DOC to the suspending solution and that active excretion of proteins, vitamins, and lipids may also occur (Aaronson, 1978). Specific compounds have been examined, including glycolate, which accounted for 13 to 92% of DO^{14}C released by a natural population of freshwater microalgae (Watt, 1966); amino acids, which accounted for 7% of extracellular DO^{14}C production in marine microalgae (Mague et al., 1980); and organic plus amino acids, which accounted for 10% of DO^{14}C released by *Ascophyllum nodosum* (Harlin and Craigie, 1975). High-molecular-weight carbohydrates can be a major component of DO^{14}C release by some phytoplankters (Myklestad et al., 1989; Lancelot, 1984; Ittekkot et al., 1981).

In one of the most comprehensive studies to date, Hellebust (1965) examined the composition of DO^{14}C released by 22 species of unicellular marine algae. He found considerable differences in the amounts and composition of the excreted material. Extracellular release ranged from 1.5 to 24% of total organic ^{14}C production in exponentially growing cultures. Glycolic acid, a product of photorespiratory activity, varied from < 1 to almost 40% of DOC release. Protein accounted for 0.2 to 5.9% of extracellular release and chloroform-soluble material for 2.8 to 10.3%. Electrodialysis was used to separate neutral, anion, and cation

Table 2.1.   Percent extracellular release (PER) and the relative contribution of low- and high-molecular-weight substances to $DO^{14}C$.

Reference	PER (%)	<700 da (% PER)	>1500 da (% PER)
Guillard and Hellebust (1971)	22	10	—
Nalewajko and Lean (1972)	4.4	20	80
	2.1	0	100
	5.1	0	60
Nalewajko and Schindler (1976)	1	8	90
	18	8	90
Steinberg (1978)	8	50	30
	4	70	20
Watanabe (1980)	42	15	60
	13	10	60
Söndergaard and Schierup (1982)	5	45	55
	4	85	15
Chrost and Faust (1983)	65	10	85
	33	50	40
	7	25	65
	14	35	60
	30	5	90
	26	15	85
	19	35	55
	7	20	75
Iturriaga and Zsolnay (1983)	8	55	35
Jensen (1983)	14	80	0
	3	55	30
	8	40	30
Cole et al. (1984)	20	—	80
Kato and Stabel (1984)	18	48	52
	26	48	52
	11	42	58
Lancelot (1984)	45	25	—
	15	26	—
	24	50	—
Hama and Handa (1987)	2	75	15
	6	65	25

After Sundh (1989).

fractions. The neutral fraction accounted for 17 to 73% of DOC released and included glycerol, proline, mannitol, glutamic acid, lysine, aspartic acid, arabinose, and glucose. The anion fraction which includes low-molecular-weight organic acids and other organic anions, accounted for 10 to 34% of DOC release. A small fraction of DOC release (2 to 6%) was found in the cation fraction.

Problems with the ^{14}C Method

The ^{14}C method has been at the center of controversy since its introduction when Steemann-Nielsen (1952) measured much lower productivities in oligotrophic ocean waters with this technique than did Riley (1938, 1939, 1941, 1944) with the O_2 light/dark bottle method. Despite a large experimental effort and widespread adoption of the ^{14}C technique, the unsatisfactory reconciliation of results from the ^{14}C method and other estimates of primary production led Peterson (1980) to review the controversies surrounding the use of the ^{14}C protocol for measuring aquatic primary production. At the time of Peterson's (1980) review it appeared that a number of problems arose from misunderstandings about algal physiology and the nature of microbial food webs in the plankton. Additional problems arose from contamination of samples with toxic trace metals such as copper, or other inhibitory substances, including plasticizers, during sample collection and incubation. The controversy was fueled by apparent discrepancies between estimates of oceanic net primary production obtained from the ^{14}C method and biogeochemical mass balances, and a number of researchers assigned inconsistencies between the results of ^{14}C-technique and other methods to problems inherent in the ^{14}C-method. The controversy has abated somewhat as intercomparisons have demonstrated that the ^{14}C method and other "bottle" techniques yield comparable rates of primary production (Williams et al, 1983; Laws et al., 1984; Bender et al., 1987), and with the appreciation that comparisons between bottle techniques and the biogeochemical approach are only valid when observations obtained on the same time and space scales are compared (Platt, 1984; Platt et al., 1984; Platt and Harrison, 1986a, b). However, reports of "enigmatic marine ecosystem metabolism" (Johnson et al., 1981) remain to be fully addressed or explained, and detailed comparisons of net gas exchange and tracer measurements of algal photosynthesis occasionally yield anomalous results (Grande et al., 1989).

Sample Contamination

The possibility of copper contamination of algal samples during photosynthesis experiments has long been recognized (Steemann-Nielsen and Brumm-Laursen, 1976), and steps to prevent contamination must be exercised whenever samples are removed from the natural environment and incubated in bottles (Carpenter and Lively, 1980; Fitzwater et al., 1982). Most of the concern about contamination has arisen when measuring phytoplankton productivity in the oligotrophic ocean regions where metal concentrations are very low (Brand et al., 1986; Gavis et al., 1981) and research vessels present a major source of contaminants. Toxic concentrations of metals can also be introduced into samples if the procedures recommended in older manuals are used. Strickland and Parsons (1972), for example, recommend cleaning glass sample bottles in chromic acid, preparing ^{14}C working solutions in a dilute salt solution, and storing the ^{14}C working solution in soft-glass ampules. All of these steps can introduce contaminating metals at concentrations that may inhibit phytoplankton photosynthesis (Fitzwater et al., 1982). Plasticizers may also inhibit algal growth and photosynthesis leading to another potential source of contamination in the tubing used to transfer aqueous samples. Latex tubing (which may be used as a spring in Van Dorn and Niskin water-bottle samplers) is particularly toxic; polyvinyl chloride tubing is generally nontoxic if thoroughly rinsed before use, but silicone tubing is considered to be the safest (Price et al., 1986).

Sources of contamination not necessarily limited to ^{14}C-productivity experiments include the sea surface microlayer in which trace elements, oils, and microorganisms can accumulate relative to the subsurface waters, the water sampling bottle, the hydrographic wire from which the water sampler is suspended, the ship itself (oxidized metal parts, paints, oils and grease, smoke, and surface-slick contaminants), any container which the sample comes in contact with, and any added reagents including the ^{14}C-bicarbonate solution. "Ultraclean" techniques have been developed by trace element chemists to avoid these sources of contamination, and the recommended procedures to obtain uncontaminated samples are summarized in Table 2.2 (Fitzwater et al., 1982).

Predicting the response of phytoplankton to trace metal additions is complicated by interspecific variability in metal tolerance and uncertainty about the concentration of free metal ions in solution. In an examination of 38 clones of marine phytoplankton, Brand et al. (1986) found that the major difference in resistance to Cu^{+2} toxicity was phylogenetic rather than geographic. Cyanobacteria were most sensitive to copper toxicity, coccolithophorids and dinoflagellates had intermediate sensitivities, and diatoms were least sensitive (Table 2.3). However, some clones from polluted estuarine waters have much greater trace metal tolerance when compared with coastal and oceanic clones (Gavis et al., 1981). It is not the concentration of total metal ions, but rather the concentration of free ions in solution that determines the availability of metals to algae (Morel et al., 1978; Petersen, 1982). Metal ions that are complexed with organic and inorganic chelating agents are not directly available to phytoplankton and thus do not affect algal growth or photosynthesis.

Nanomolar concentrations of cupric ion inhibited the growth of the diatom *Thalassiosira pseudonana* (Sunda and Guillard, 1976) and the motility of the dinoflagellate *Gonyaulax tamarensis* (Anderson and Morel, 1978).

Table 2.2. "Ultraclean" techniques for primary productivity experiments recommended by Fitzwater et al. (1982).

1. Collect samples in nonmetallic samplers. All metal parts must be coated in plastic or replaced by nonmetallic parts. Black rubber parts must be replaced with silicone parts or enclosed within silicone or polyethylene tubing. Teflon coated PVC Go-Flo ball-valve samplers (General Oceanics) were recommended.

2. Samplers must be scrupulously clean. External surfaces should be cleaned with a laboratory detergent. Internal surfaces should be cleaned by sequential soakings for 1 week in Micro (American Scientific Products), 0.5molm^{-3} glass redistilled nitric acid, 0.25N quartz-distilled nitric acid and copiously rinsed with ultrapure (Milli-Q type, or similar) water.

3. Metal hydrowires should be replaced by a synthetic hydroline spooled onto plastic-wrapped hydrodrum, and solid teflon or PVC messengers should be used.

4. Samples should be handled in an area that is free from airborne particles in a self-contained positive pressure laboratory or flow cabinet provided with a filtered air supply. The sample handling area should be constructed of wood and plastic.

5. Polycarbonate* bottles should be used for incubating samples. Bottles should be soaked in Micro for 5 days, rinsed with Milli-Q water, soaked in 0.25N nitric acid for 3 days, and rinsed with Milli-Q water.

6. Working ^{14}C-bicarbonate solutions should be prepared from high activity (mCi·mmol^{-1}) manufacturer's stocks in 0.3 g·dm^{-3} sodium carbonate solutions. Working solutions should be stored in acid-cleaned Teflon bottles.

7. All pipette tips used to dispense ^{14}C solutions should be rinsed three times in 2molm^{-3} redistilled nitric acid, and six times with Milli-Q water prior to use. Small additions (0.25 cm^{-3}) of ^{14}C stock solutions are used to further reduce potential contamination.

8. Plastic gloves and lab coats should be worn at all times when injecting samples.

*Strong acids which modify the surface characteristics of polycarbonate bottles should be avoided.

In a comprehensive study of 24 clones of marine phytoplankton, Gavis et al. (1981) found that free cupric ion concentrations of 1 nM to 1 pM (0.06–60 pg Cu·dm^{-3}) can inhibit phytoplankton growth. By way of comparison, an analysis of potential contamination associated with each component of the standard ^{14}C technique led Fitzwater et al. (1982) to

Table 2.3.   Copper sensitivity of phytoplankton growth.

	pCu for 50% Inhibition of Growth	
	Range	Mean ± Standard Deviation
Diatoms	9.45–10.47	10.04 ± 0.29
Coccolithophorids	9.5–10.65	10.43 ± 0.24
Dinoflagellates	9.82–11.07	10.40 ± 0.44
Cyanobacteria	10.25–11.54	10.88 ± 0.44

Adapted from Brand et al. (1986).

conclude that additions of copper (in pg) were possible if "ultraclean" techniques were not used.

The extent of inhibition of photosynthesis by metal additions will depend on the sensitivity of phytoplankton present in a water sample to these metals, the extent of contamination during sample processing and the free metal ion buffering capacity of the water sample. There have been few comparisons of ^{14}C production using ultraclean and less rigorous techniques, and, as might be expected, the results have been variable, with some investigators indicating no inhibition and others a slight inhibition of phytoplankton photosynthesis. Fitzwater et al. (1982) report inhibition of ^{14}C uptake by a natural phytoplankton sample when copper additions exceeded 0.1 pg $Cu \cdot dm^{-3}$, with a 50% reduction in the ^{14}C-uptake rate when the addition reached 2 pg $Cu \cdot dm^{-3}$. However, Marra and Heinemann (1984) found no difference in ^{14}C uptake between ultraclean and standard techniques in three different locations.

Recent observations of primary productivity in the oligotrophic North Pacific subtropical gyre have employed ultraclean ^{14}C techniques. These observations have yielded ^{14}C uptake rates about twice as high as the historical average for the same location (Laws et al., 1987; Marra and Heinemann, 1987). Not only were the areal rates of primary production higher in these experiments, but the chlorophyll-specific light-saturated photosynthesis rates were higher as well. For example, in an early study by Eppley et al. (1973) a maximum P_{chl} of approximately 30 g $C \cdot g^{-1}$ $chla \cdot d^{-1}$ was found, whereas more recently Laws et al. (1989) obtained maximum P_{chl} of 100–150 g $C \cdot g^{-1}$ $chla \cdot^{-1} d^{-1}$. Since P_{chl} tends to decrease when microalgae are exposed to adverse growth conditions, it would appear that contamination may have been a problem in some of the earlier investigations of open ocean primary productivity.

What Does ^{14}C Uptake Measure?

The preceding section considered problems that may arise in any study of algal photosynthesis in which samples for observation are enclosed. These potential "housekeeping" problems can be overcome by careful techniques. One problem that has plagued the application of the ^{14}C technique since its introduction to photosynthesis research is "what does ^{14}C assimilation actually measure?" Methodological or interpretative problems unique to the ^{14}C measurements are considered in this section. These include a lack of specificity for photosynthetic carbon fixation and an inability to directly measure algal respiration. The first of these problems can be examined, in part, through a comparison of dark-versus-light $^{14}CO_2$ fixation. The problem of measuring algal respiration requires a consideration of isotope equilibration among intracellular carbon pools.

Dark versus Light CO_2 Fixation

A practical problem faced by ecologists and physiologists in applying the ^{14}C method is how to use the results from dark-bottle incubations to "correct" the light-bottle data. Various practices have been applied, including subtracting the dark-bottle value from the light-bottle value in direct analogy with the treatment of light and dark measurements in net gas exchange determinations, ignoring the dark-bottle value and calculating carbon production only on the basis of the light-bottle results, or subtracting some fraction of the dark-bottle results, usually 50%, from the light-bottle value (Peterson, 1980). Before addressing these questions, we need to consider the processes involved in dark ^{14}C accumulation.

Net inorganic carbon fixation by algae, as in all plants, involves carboxylation of ribulose bisphosphate (RuBP) to form 3-phosphoglycerate by the enzyme ribulose 1–5-bisphosphate carboxylase/oxygenase (RUBISCO) in what is called C_3 photosynthesis. Carbon fixation by RUBISCO in the photosynthetic carbon reduction cycle (PCRC) is autocatalytic. That is, the substrate RuBP, which is consumed in the carboxylation step, is also regenerated at the end of the cycle with a net increase in the concentration of intermediates of the cycle during photosynthesis. These intermediates can be used to support an enhanced rate of PCRC activity or can be removed from the cycle for biomass accumulation (Fig. 2.1). Biomass accumulation via the PCRC cycle can normally only proceed under illuminated conditions, or within minutes following a transition from light to darkness, because of the requirement for light-generated ATP and reductant for CO_2 assimilation.

In some vascular plants, inorganic carbon is initially fixed into a four-carbon acid via a β-carboxylation step prior to decarboxylation and refixation of CO_2 at a different site within the leaf. This pattern of carbon fixation is referred to as C_4 photosynthesis. Rapid labeling of organic acids via β-carboxylations in algae was initially interpreted as indicative of a C_4-like metabolism (Beardall et al., 1976). The C_4 pathway involves an initial carbon fixation via a β-carboxylase, with subsequent decarboxylation of a C_4 acid, which provides CO_2 that is refixed by RUBISCO. However, in actively photosynthesizing algae the formation of malate and aspartate is preceded by fixation of CO_2 into 3-phosphoglyceric acid (3-PGA) by RUBISCO (i.e., by a C_3 mechanism) (Kremer, 1981), and recent experiments with microalgae (Burns and Beardall, 1987) and macroalgae (Bidwell and McLachlan, 1985) indicate that these organisms do not have a C_4 mechanism. In algae, β-carboxylations do not appear to be

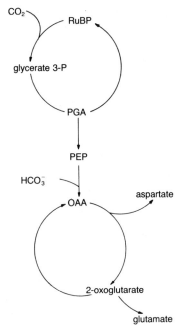

Figure 2.1. Coupling of photosynthetic carbon reduction cycle (PCRC) and tricarboxylic acid cycle (TCA) illustrating major routes of CO_2 and HCO_3^- incorporation by algae. Inorganic CO_2 is fixed primarily by carboxylation of ribulose bisphosphate (RuBP) to form glycerrate 3-P. Phosphoglyceric acid (PGA) can be used to regenerate RuBP, diverted from the PCRC to synthesize storage carbohydrates and/or diverted from the PCRC to form phosphoenol pyruvate (PEP). Inorganic HCO_3^- is fixed primarily by β-carboxylation of PEP to form oxaloacetate (OAA) which enters the TCA cycle, or serves as a precursor for the synthesis of aspartate and related amino acids. Synthesis of glutamate and related amino acids occurs via 2-oxoglutarate, an intermediate of the TCA cycle. Modified from Raven (1984) and Mortain-Bertrand et al. (1988)

involved in photosynthetic carbon fixation but rather in the coupling of the first stable product of photosynthesis (3-PGA) to protein and lipid synthesis (Mortain-Bertrand et al., 1988) through regeneration of Krebs tricarboxylic acid (TCA) cycle intermediates (Fig. 2.1). Replenishment of TCA cycle intermediates also requires inorganic carbon fixation, but does not result in net carbon assimilation. Anapleurotic mechanisms that involve β-carboxylations are catalyzed by the enzymes pyruvate carboxylase, phosphoenolpyruvate (PEP) carboxylase, and phosphoenolpyruvate carboxykinase (Appleby et al., 1980). Because the demand for intermediates is likely to continue at the same rates in the light and darkness (Raven, 1984), it is not surprising that the rate of β-carboxylation is often observed to be the same in darkened and illuminated algae (Kremer, 1981). However, this may not always be the case. Mortain-Bertrand et al. (1987) demonstrated differences in the rate of β-carboxyla-

Table 2.4.   Ratio of β-carboxylation to carboxylation by the C_3 pathway in *Skeletonema costatum* cultured on light:dark cycles and under continuous illumination.

	β-*carboxylation*/C_3 *carboxylation*	
	light period	dark period
Continuous light	0.12	—
12 h:12 h	0.08	0.009–0.017
2 h:2 h	0.03	0.12

Adapted from Mortain-Bertrand et al. (1987).

tion in the light and darkness in the marine diatom *Skeletonema costatum* (Table 2.4). Johnston and Raven (1989) also report higher rates of β-carboxylation in the light in the macroalga *Ascophyllum nodosum*.

One measure of the importance of β-carboxylation relative to carboxylation by RUBISCO is the ratio of the rates of ^{14}C assimilation in darkened and illuminated cells. In actively growing algae, the rate of dark $^{14}CO_2$ fixation is typically < 5% of the light-saturated rate of $^{14}CO_2$ fixation. However, dark:light carboxylation ratios for algae range from less than 0.01 to over 0.4 (Mortain-Bertrand et al., 1988). Where phenotypic variation has been investigated, the dark:light carboxylation ratio has been observed to increase under adverse growth conditions (Descolas-Gros and Fontugne, 1985), perhaps contributing to the reported variability. A recent compilation of ratios of RUBISCO to phosphoenol pyruvate carboxylase activities in microalgae (Glover, 1985) indicates a range from 0.5 to 29.2, consistent with the wide range of dark-to-light ^{14}C fixation ratios reported by Mortain-Bertrand et al. (1988).

The TCA cycle is a universal feature of heterotrophic and autotrophic organisms, and β-carboxylation by heterotrophic organisms typically proceeds at approximately 5% of the growth rate, although a range of < 1 to 30% has been reported (Li, 1982). The ratio of dark:light carboxylation in the euphotic zone of oceans and lakes varies from < 0.01 to 0.5 (Morris et al., 1971; Hecky and Fee, 1981; Taguchi, 1983), a range not greatly different from that reported for algae (Mortain-Bertrand et al., 1988). High ratios in microplankton samples indicate either that the phytoplankton are growing under suboptimal conditions or that heterotrophic organisms make a significant contribution to β-carboxylations.

Addition of inorganic nitrogen (ammonium) to nitrogen-limited microalgae is known to stimulate TCA cycle activity and thus increase the β-carboxylation rate. The enhancement ratio, defined as the rate of dark ^{14}C fixation in the presence of added ammonium divided by the rate in a control without added ammonium, has been suggested as an indicator of the degree of nitrogen limitation (Morris et al., 1971; Yentsch et al.,

1977). In the diatom *Chaetoceros simplex*, the enhancement ratio increases with C:N ratio, and by implication with increases in the degree of nitrogen limitation (Goldman and Dennett, 1986). This increase in the enhancement ratio is accompanied by a proportional increase in the ratio of dark-enhanced ^{14}C accumulation to ^{14}C uptake in the light. The enhancement ratio increased from 1 to 30, and the ratio of dark:light carboxylation increased from < 0.01 to 0.35 with increased nitrogen limitation in *Chaetoceros simplex* (Goldman and Dennett, 1986). This range of dark:light carboxylation ratios is almost as great as that reported from interspecific comparisons (Mortain-Bertrand et al., 1988).

If replenishment of the TCA cycle intermediates is the major process accounting for dark ^{14}C uptake and the rate of β-carboxylation is similar in the light and darkness, then it would seem appropriate to subtract the dark-bottle from the light-bottle value. The difference between light-bottle and dark-bottle ^{14}C assimilation rates would presumably equal the rate of CO_2 fixation by RUBISCO. However, some investigators have suggested that subtracting the dark-bottle ^{14}C uptake from the light-bottle uptake may lead to serious underestimates of photosynthetic carbon fixation because light inhibits some or all of the nonphotosynthetic organic ^{14}C accumulation (Morris et al., 1971). Legendre et al. (1983) suggested that a proper correction for nonphotosynthetic ^{14}C accumulation was provided by dichlorophenyl dimethylurea (DCMU or diuron) poisoned samples, but this awaits further verification.

Isotope Equilibration

Interpretation of the ^{14}C-uptake experiments requires a compartmental model of algal metabolism (Fig. 2.1). Such a model consists of a set of equations describing the compartments within which radioactive tracer accumulates together with the rules for evaluating the rates of transfer between these compartments (Smith and Horner, 1981). Phytoplankton productivity experiments have most often been interpreted in terms of a simple model (Fig. 2.2) consisting of two compartments, an extracellular pool of inorganic carbon, and a homogeneous intracellular pool of organic carbon (Peterson, 1980; Dring and Jewson, 1982).

The homogeneous intracellular pool model was first treated explicitly by Hobson et al. (1976) and was considered in some detail by Dring and Jewson (1982). The model can be solved analytically, leading to the following equations for particulate organic carbon (POC) concentration, particulate organic ^{14}C activity ($PO^{14}C$), and specific activity of the POC ($PO^{14}C/POC$).

Model 1

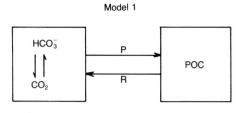

Model 2

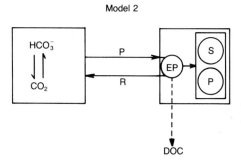

Figure 2.2. Typical compartmental models of inorganic carbon metabolism used in the kinetic analysis of ^{14}C uptake kinetics. Model 1 is the traditional two-compartment model treated by Hobson et al. (1976) and Dring and Jewson (1982) in which photosynthesis (P) and respiration (R) mediate the exchange between an external pool of CO_2 (HCO_3^-) and a particulate organic carbon (POC) pool. Model 2 is the four-compartment model treated by Smith and Platt (1984) which includes an exchanging pool (EP) excretion of dissolved organic carbon (DOC) and synthesis of storage carbohydrates (S) and functional polymers (P).

$$POC(t) = POC(0) \exp (P_g - R)t \qquad (2.2)$$

$$PO^{12}C(t) = [POC(0) / (1 + TCO_2^*)] \qquad (2.3)$$
$$[\exp (P_g - R)t + TCO_2^* \exp (-Rt)]$$

$$PO^{14}C(t) = [TCO_2^* POC(0) / (1 + TCO_2^*)] \qquad (2.4)$$
$$[\exp (P_g - R)t - \exp (-Rt)]$$

$$PO^{14}C/POC = \frac{TCO_2^* [\exp (P_g - R)\,t - \exp (-Rt)]}{\exp(P_g - R)t + TCO_2^* \exp (-Rt)]} \qquad (2.5)$$

where TCO_2^* is the specific activity of inorganic ^{14}C, P_g is the photosynthesis rate (h^{-1}), and R is the respiration rate (h^{-1}). The homogeneous, intracellular-pool model has been the focus of theoretical studies of the ^{14}C method largely because of its analytical tractability. However, application of this model to $PO^{14}C$ time series usually leads to unrealistically high estimates of P, R, and R/P_g (Dring and Jewson, 1982; Li and Harrison, 1982). Dring and Jewson (1982) recognized this limitation and proposed several modifications to the model structure, including the recognition of heterogeneity in the intracellular carbon pool. However, the more complicated models cannot be solved analytically, and interpreting $PO^{14}C$ time series in terms of these more complicated models requires a compartmental analysis of tracer kinetics (Smith and Horner, 1981).

Smith and Platt (1984) showed that a model with two intracellular pools, one exchanging and one synthetic, is required to describe $PO^{14}C$ accumulation and $DO^{14}C$ release by phytoplankton (Fig. 2.2). The exchanging pool is small (approximately 10% of total cell carbon) and rapidly cycling relative to total cell carbon (Smith and Platt, 1984; Smith and Geider, 1985). Respiration rates of illuminated cells can be calculated as part of the curve-fitting procedure (Smith and Platt, 1984). These rates are uniformly low (< 0.1 d^{-1}) and independent of growth rate for microalgae in balanced growth (Smith and Geider, 1985). In contrast, a wider range (< 0.1–0.6 d^{-1}) and also a growth-rate dependence, is observed in direct measurements of microalgal respiration (Geider and Osborne, 1989). Estimation of the size of the exchanging pool, which was assumed to be the site for both excretion and respiration, relied heavily on time-series measurements of $DO^{14}C$ activity, which showed more pronounced nonlinearity than did $PO^{14}C$ activity (Smith and Platt, 1984; Smith and Geider, 1985). It is likely that excretion and respiration do not occur from the same compartment, and thus Smith and Geider's (1985) estimates of respiration may be in error.

The specific activity of the intracellular inorganic carbon pool has been assumed to equal that of the extracellular TCO_2 pool in all of the theoretical models considered so far. However, this assumption is not necessarily valid, posing another problem in the interpretation of ^{14}C assimilation experiments. Differences between the intracellular and extracellular specific activity of the inorganic ^{14}C may arise from intracellular production of unlabeled CO_2 due to respiration of "old" carbon, which effectively dilutes the added $T^{14}CO_2$. Refixation of this $^{12}CO_2$ will lead to underestimates of photosynthesis rates in ^{14}C experiments.

In an attempt to pigeonhole observations into neat categories, researchers have tried to determine the conditions under which the ^{14}C method measures net or gross photosynthesis. In any model of intracellular carbon pools organic ^{14}C accumulation will equal gross photosynthesis (P_g) when the $^{14}C/^{12}C$ ratio within the exchanging pool is infinitely less than the $^{14}C/^{12}C$ in the inorganic pool. Likewise, $PO^{14}C$ accumulation will equal P_n when the exchanging and inorganic pools have equal $^{14}C/^{12}C$ ratios. In most photosynthesis experiments, something between gross and net photosynthesis is likely to be measured because the $^{14}C/^{12}C$ ratio in the exchanging pool will neither be infinitely small, nor equal to the $^{14}C/^{12}C$ ratio of inorganic carbon. When the $P_g{:}R$ ratio is high $[(P_g - R)/P \approx 1]$, the variable extent of isotope equilibration does not present a major problem because $P_n \approx P_g$. However, there may be difficulty in assigning ^{14}C accumulation to P_n or P_g when the ratio of photosynthesis to respiration is low. This is likely to occur under conditions of light limitation when the irradiance limits the rate of photosynthesis or under nutrient limitation when the capacity for photosynthesis diminishes relative to the capacity for respiration.

A potential problem in ecological investigations which may be more important than intracellular isotope disequilibrium is the effect of isotope exchange between algae and other organisms during productivity experiments with natural samples. Bacteria can assimilate and respire dissolved organic ^{14}C, and protozoan or small metazoans included with the sample may graze on particulate organic ^{14}C of algal origin.

Zooplankton Grazing and ^{14}C Assimilation

Most theoretical studies of the relationship between biomass accumulation and ^{14}C-tracer dynamics have been limited to temporal changes in the specific activity of algal cells (Hobson et al., 1976; Buckingham, et al., 1975; Dring and Jewson, 1982), as discussed above. However, under natural conditions, it is likely that the transfer of carbon within food webs and the labeling of microzooplankton and bacteria with ^{14}C will also have a pronounced effect on ^{14}C-tracer dynamics. Recent studies have approached this problem, including those of Jackson (1983), Smith et al. (1984), and Marra et al. (1988); because all three models have many features in common, this discussion focuses on Jackson's model.

Jackson (1983) examined the effects of zooplankton grazing on phytoplankton growth, community metabolism, and ^{14}C assimilation into organic matter. Phytoplankton biomass was assumed to increase during the day when photosynthetic carbon fixation exceeds losses due to microzooplankton grazers. Phytoplankton respiration was assumed to be negligible and losses of phytoplankton biomass were attributed exclusively to the grazing activity of the microzooplankton. In the steady-state system postulated by Jackson, phytoplankton biomass decreased at night by an amount equal in magnitude to the daytime increase. The model was expressed in terms of Eq. 2.6–2.8 for total carbon fluxes and Eqs. 2.9–2.11 for ^{14}C fluxes.

$$dB \, / \, dt = (\mu - g)B = (1 - f)\mu B \qquad \text{during daylight} \qquad (2.6)$$

$$dB \, / \, dt = -gB = -f\mu B \qquad \text{during darkness} \qquad (2.7)$$

$$dZ \, / \, dt = ef\mu B \qquad \text{at all times} \qquad (2.8)$$

where B is the phytoplankton concentration ($g \cdot m^{-3}$), Z is the zooplankton concentration ($g \cdot m^{-3}$), μ is the phytoplankton growth rate (d^{-1}), g is the specific rate, f is the specific grazing rate as a fraction of the specific growth rate, and e is the conversion efficiency of plant carbon into zooplankton carbon.

$$dB^* \, / \, dt = TCO_2^* \, \mu B - f\mu B^* \qquad \text{during daylight} \qquad (2.9)$$

$$dB^* \, / \, dt = -f\mu B^* \qquad \text{during darkness} \qquad (2.10)$$

$$dZ^* \, / \, dt = e\mu f B^* \qquad \text{at all times} \qquad (2.11)$$

where B^* is the ^{14}C activity in phytoplankton, Z^* is the ^{14}C activity in zooplankton, and TCO_2^* is the specific activity of the dissolved inorganic carbon. The quantity measured at the end of an incubation is the total organic ^{14}C collected on a filter, given by

$$C^* = B^* + Z^* \qquad (2.12)$$

The results from Jackson's model with $\mu = 2 \, d^{-1}$, $f = 0.5$, and $e = 0.37$ for a 12-hr photoperiod, followed by 12 hr darkness, are illustrated in Fig. 2.3. Phytoplankton abundance increases by approximately 60% during the day and decreases by an equal amount at night. In the absence of grazing, the daytime increase would have been 170%, giving a phytoplankton concentration of exp (1.0) = 2.7 times the initial value. Thus, a significant fraction of net photosynthesis is assumed to be simultaneously consumed by microzooplankton grazers, limiting the rate of increase of phytoplankton biomass. Phytoplankton become labeled to a specific activity equal to 0.63 (i.e., 1.7/2.7) of the specific activity of the TCO_2 pool after 12 hr. In this model, the specific activity of the particulate carbon remineralized by microzooplankton is assumed to equal the specific activity of phytoplankton carbon. Thus, even after 12 hr in the light, the instantaneous loss rate of ^{14}C from the particulate pool will only equal 63% of the rate of the true rate of loss of particulate organic carbon. Only when the phytoplankton become labeled to the same specific activity as the inorganic carbon pool will instantaneous loss rates for $PO^{14}C$ equal loss rates of POC.

According to Jackson's (1983) model, the incorporation of ^{14}C into particulate matter results in a slight overestimate of the rate of change in

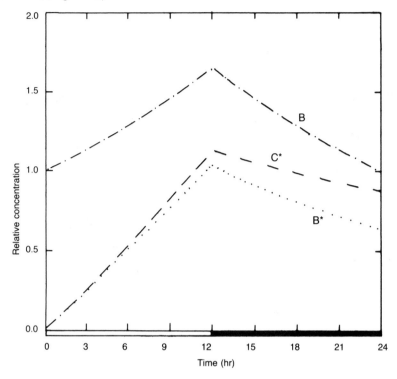

Fig. 2.3. Daily variation in total phytoplankton carbon (B), radioactively-labeled phytoplankton carbon (B*) and total radioactively-labeled carbon (C*) for a 12-hr light period followed by a 12-hr dark period. The difference between the total radioactively labeled carbon and the total phytoplankton carbon is due to carbon present in the zooplankton. After Jackson (1983). See text for details.

phytoplankton biomass, because some label accumulates in the microzooplankton. This activity is indicated by the difference between B^* and C^* in Fig. 2.3. The extent to which phytoplankton production is overestimated by particulate ^{14}C accumulation depends on the phytoplankton growth rate, the incubation time and the grazing pressure. The accumulation of ^{14}C in microzooplankton will be significant under conditions of high phytoplankton growth and microzooplankton grazing rates (Jackson, 1983).

One important generalization from Jackson's (1983) theoretical study is that the product of the growth rate and the incubation time (μT) determines the extent to which deviations in $PO^{14}C$ accumulation from phytoplankton production are likely to arise. The dimensionless parameters μT is a measure of how much the system has turned over during the incubation. Unfortunately, the estimate of μ is problematic, but it is possible to gain some insight into the appropriateness of ^{14}C experiments by assuming upper and lower limits to μ. In general, there is a greater chance that ecological interactions will affect microbial populations and ^{14}C accumulation as μT increases. As a consequence, Jackson (1983) recommended short incubation times to avoid population changes in incubation vessels during productivity experiments.

Comparison of Net Gas Exchange and ^{14}C Assimilation

The photosynthetic quotient (PQ) for net gas exchange was defined in Chapter 1 as the ratio of O_2 evolution to TCO_2 assimilation. An analogous quotient, referred to as the *apparent* PQ can be defined as the ratio of gross O_2 evolution to ^{14}C assimilation. Apparent PQs are defined in terms of gross oxygen evolution because of uncertainty over the extent to which heterotrophic organisms affect net gas exchange rates in natural waters. Apparent PQs range from 1–3 mol $O_2 \cdot mol^{-1} CO_2$. An apparent PQ of 1.95 (SD = 0.36; range 1.5–2.7) was obtained during a 19-d experiment in which NO_3 decreased from 20 μM to undetectable concentrations in a large volume plastic sphere incubated in coastal seawater (McAllister et al., 1961). Jewson (1977) observed an apparent PQ of 1.0 for phytoplankton from Lough Neagh, Northern Ireland and Grande et al. (1989) found apparent PQs of approximately unity in pure cultures. The slope of a plot of O_2 evolution versus ^{14}C fixation yielded an apparent PQ of 1.75 for phytoplankton from Lake Kinneret (Megard et al., 1985); however, the apparent PQ declined markedly at light levels that inhibited O_2 evolution but had little effect on rates of organic ^{14}C production. Williams et al. (1983) found a mean apparent PQ of 1.35 (range 0.92–1.66) for phytoplankton from the surface mixed layer of the oligotrophic North Pacific gyre. Platt et al. (1987) measured apparent PQs that ranged from 1.1 to 2.7 for phytoplankton from the Canadian Arctic.

Comparisons of ^{14}C incorporation and O_2 evolution have also been made on macrophytes. Hoffman and Dawes (1980) obtained an apparent PQ of 1.2–1.4 for two red macroalgae, and Jensen et al. (1985) obtained values of approximately unity for four deep-water calcareous *Halimeda* species when organic ^{14}C production was compared with O_2 evolution in plants incubated at an irradiance of 20 $\mu mol \cdot m^{-2} \cdot s^{-1}$.

Variability in apparent PQs is likely to arise from many of the same processes that affect the PQs for net gas exchange (see Chapter 1). The assimilation of inorganic nitrogen appears to be a major factor affecting apparent PQs in marine phytoplankton. Williams et al. (1979) and Raine (1983) found apparent photosynthetic quotients which ranged from 1.05 to 2.25 in coastal phytoplankton and outdoor algal cultures. The photosynthetic quotient was 1.25 when ammonium was the nitrogen source (Williams et al., 1979; Raine, 1983). However, when nitrate was the nitrogen source, the rate of oxygen evolution depended on both carbon fixation and nitrate reduction, with a stoichiometric relation given by Eq. 2.13.

$$\delta O_2/\delta t = 1.25\ \delta TCO_2/\delta t + 2\delta NO_3/\delta t \tag{2.13}$$

Thus, variability in the photosynthetic quotient appears to arise largely from changes in the relative rates of NO_3 and CO_2 reduction.

$$PQ = \delta O_2/\delta TCO_2 = 1.25 + 2\delta NO_3/\delta TCO_2 \tag{2.14}$$

However, other factors may contribute to variations in apparent photosynthetic quotients. These include a failure to account for all ^{14}C fixation into organic matter if dissolved organic ^{14}C and volatile organic ^{14}C are not recovered.

Modifications to the [14]C Method

In addition to the increase in sensitivity over net-gas-exchange techniques, the ^{14}C method offers other advantages, including the ability to separately measure the production of DO[14]C, PO[14]C, and carbonates that may be accumulated by certain algae (Paasche, 1963). This technique can also be used to measure the production of particular organic molecules or classes of macromolecules and to estimate the photosynthesis rates of individual microalgal cells.

Biochemical Fractionation of Organic [14]C

Biochemical fractionation of assimilated ^{14}C was introduced to phytoplankton research by Morris et al. (1974), based on modifications of procedures developed for *Escherichia coli* (Roberts et al., 1963). Separation of biochemical fractions usually begins with chloroform–methanol (2:1 by volume) extraction, followed by acid hydrolysis in hot ($> 80°C$) trichloracetic acid (TCA) (Li et al., 1980). Water is added to the chloroform–methanol extract to obtain an aqueous phase that contains low-molecular-weight molecules and a chloroform phase that contains lipids. Proteins are obtained in the TCA-insoluble fraction, whereas carbohydrates are obtained in the TCA-soluble material. None of these fractions are biochemically "pure." The carbohydrate phase contains nucleic acids, and a significant fraction of reserve carbohydrates may be extracted with chloroform–methanol and thus appear in the metabolite pool (McConville et al., 1985, cited in Hama et al., 1988). In addition, protein cannot be completely separated from polysaccharide by a TCA extraction (Hama and Handa, 1987), and diatom chitin remains in the TCA insoluble fraction as a possible contaminant (Smucker and Dauso, 1986). Although based on the properties of different macromolecular classes, the simple extraction protocol outlined above is strictly empirical, and changes in protocol or species may significantly influence the results (Hitchcock, 1983). Nevertheless, this approach provides additional information, which has been effectively utilized in studies of algal photosynthesis.

Application of biochemical fractionation has been most useful in evaluating the rate of protein synthesis (incorporation of ^{14}C into the TCA-insoluble fraction). In unialgal cultures, under controlled conditions, the ratio of protein synthesis to total carbon assimilation covaries with the nutrient-limited growth rate. For example, Smith and Geider (1985) found that the TCA-insoluble activity increased from 30% to 50% of total carbon assimilation in *Thalassiosira pseudonana* over a range of nutrient-limited relative growth rates (μ/μ_m; where μ is the observed growth rate and μ_m is the potential resource unlimited growth rate) from 0.15 to 1.0. The upper limit is consistent with the rule of thumb approximation that protein accounts for 50% of microalgal carbon at μ_m. Under light-limited conditions the TCA-insoluble fraction accounted for an average of 59% of total carbon assimilation, independent of growth rate in *Thalassiosira pseudonana* (Smith and Geider, 1985). Using a similar fractionation protocol, DiTullio and Laws (1986) found that TCA-insoluble activity accounted for 46% ± 6% of cell carbon at μ_m in five species of marine phytoplankton, decreasing to 15% ± 6% as μ/μ_m approached zero. Using a different fractionation protocol, Lohrenz and Taylor (1987) found much lower percentages of cell carbon in the TCA-insoluble fraction: these ranged from only 10% to 20% of total carbon assimilation in *Nannochloris atomus* over a range (0.2–1.0) of nutrient-limited relative growth rates. Both the fractionation protocol and the species under investigation appear to influence the relationship between percentage ^{14}C activity in the TCA-insoluble fraction and μ/μ_m under nutrient-limiting conditions.

DiTullio and Laws (1986) used the percentage activity in the TCA-insoluble fraction as a measure of the relative growth rate of marine phytoplankton. Measurement of ^{14}C activity in the TCA-insoluble fraction also allows a direct evaluation of phytoplankton protein synthesis rate that can be compared with total microbial protein synthesis by measurement of sulfur or nitrogen assimilation (DiTullio and Laws, 1984) in order to assess the importance of phytoplankton in overall community metabolism. This approach is subject to considerable uncertainty due to interspecific differences in the relationship between protein content and μ/μ_m, and the fractionation efficiency outlined above. A potentially more accurate method for evaluating rates of carbon assimilation into protein and carbohydrates employs the stable isotope ^{13}C and is described later in this chapter (see pp. 61 to 62).

Chlorophyll Labeling

Redalje and Laws (1981) introduced a method for estimating phytoplankton growth rate by measuring the specific activity of ^{14}C in chlorophyll *a*. This procedure largely overcomes the problem of having to estimate a carbon-to-chlorophyll *a* ratio before calculating a growth rate from measurements of ^{14}C incorporation and chlorophyll *a* concentration (Eppley, 1968). Chlorophyll *a* is an ideal molecular marker to examine in labeling studies, because it is common to all oxygenic photoautotrophs, is readily quantified by fluorescence or absorption spectra, and can be isolated by thin-layer or high-performance liquid chromatography (see Chapter 4). The rate of chlorophyll *a* synthesis is given by the amount of ^{14}C in chlorophyll *a* divided by the product of the chlorophyll *a* concentration, the specific activity of inorganic carbon, and the duration of the experiment.

$$\rho_{chl} = 0.7399 \frac{^{14}C - chla}{[chla] \, \delta t \, TCO_2^*} \tag{2.15}$$

where ρ_{chl} is the rate of chlorophyll synthesis, 0.7399 is the carbon-to-chlorophyll *a* ratio of the chlorophyll *a* molecule. ^{14}C$-$chl*a* is the activity in the chlorophyll *a* molecules, TCO_2^* is the specific activity of dissolved inorganic ^{14}C, [chl*a*] is the chlorophyll concentration, and δt is the duration of the experiment. This equation assumes that the specific activity of the precursors of chlorophyll *a* are identical to that of the inorganic carbon pool.

Calculating the growth rate from ρ_{chl} requires the additional assumption that the specific activity of the carbon in chlorophyll *a* is equal to that of total phytoplankton carbon at the end of the experiment. This assumption has been tested and found to be accurate in a limited number of laboratory investigations using phytoplankton in balanced growth (Redalje and Laws, 1981; Redalje, 1983; Welschmeyer and Lorenzen, 1984). However, it may not be accurate under all conditions. In particular, rates of chlorophyll *a* synthesis and total carbon fixation may vary immediately following a change in environmental conditions. Reductions in irradiance or increases in nutrient availability will favor chlorophyll synthesis over total carbon accumulation ($\rho_{chl} > \mu_C$) whereas an increase in light level or a decrease in nutrient availability would favor the opposite ($\mu_C > \rho_{chl}$) (Geider and Platt, 1986). Significantly, Welschmeyer and

Lorenzen (1984) calculated that phytoplankton carbon-to-chlorophyll *a* ratios increased during chlorophyll-*a*-labeling experiments, suggesting that cells were acclimating to either increased irradiance or decreased nutrient availability during confinement at fixed depths. An increased carbon-to-chlorophyll *a* ratio may result from either decreases in the rate of chlorophyll synthesis relative to the carbon accumulation rate or increases in the degradation rate of chlorophyll, even under conditions of rapid chlorophyll *a* synthesis. There is conflicting information on rates of chlorophyll turnover in algae with some investigations favoring high rates (Riper et al., 1979) and others advocating a much more conservative behavior (Redalje and Laws, 1981; Welschmeyer and Lorenzen, 1984). Despite potential limitations, the chlorophyll-*a*-labeling technique appears to provide one of the most reliable methods of measuring microalgal growth rates. It has been applied in a number of investigations (Gieskes and Kraay, 1986; Laws et al., 1987) and has recently been extended to investigate labeling of taxon-specific carotenoids (Gieskes and Kraay, 1989).

Single Cell Rates of Photosynthesis of Microalgae

Two approaches have been applied to determine the photosynthesis rate of individual microalgal cells. Assimilation of ^{14}C can be determined either by microautoradiography (Knoechel and Kalff, 1978) or by liquid scintillation counting of individual cells (Rivkin and Seliger, 1981). These approaches allow the determination of photosynthesis rates of individual microalgae in mixed populations, but both techniques are selective in that only a portion of the total number of species present can be examined. Although techniques based on isolating individual cells can provide a quantitative measure of carbon assimilation, they can only be readily applied to large cells, whereas most planktonic environments are dominated by small phytoplankton. For example, in the Sargasso Sea and in the Gulf Stream, the large dinoflagellate *Pyrocystis noctiluca*, which was amenable to single-cell isolation, accounted for only 0.02% to 0.07% of the total ^{14}C assimilation (Rivkin et al., 1982; 1984). Microautoradiography, in contrast, can be applied to the smallest algae and cyanobacteria (Douglas, 1984; Iturriaga and Marra, 1988) but is largely a qualitative technique because of losses of ^{14}C activity during fixation and sample preparation (Silver and Davoll, 1978) and uncertainty in the quantification of ^{14}C activity (Knoechel and Kalff, 1976a). Despite these limitations, both techniques can provide insights into the species-specific responses of microalgae to environmental variables.

Microautoradiography

At least three laboratories simultaneously and independently adapted microautoradiography to the measurement of phytoplankton productivity (Maguire and Neill, 1971; Watt, 1971; Stull et al., 1973). In a typical protocol, a sample is collected and incubated in the same manner as for a normal ^{14}C experiment. At the end of the incubation, the microalgae are collected by filtration onto a polycarbonate sieve, the sample is transferred to a gelatin film on a microscope slide, and the slide is dipped into a photographic emulsion (Tabor and Neihof, 1982). Radioactive decay of ^{14}C leads to the production of silver grains, the density of which can be used to quantify photosynthesis rates. Use of a thick emulsion allows "tracks" of silver grains to be identified and has been suggested to provide a more accurate measure of cell-specific ^{14}C content (Knoechel and Kalff, 1976b), although with adequate precautions, grain density autoradiography can provide equivalent accuracy (Pearl and Stull, 1979; Davenport and Maguire, 1984).

Silver and Davoll (1978) considered the loss of ^{14}C activity that accompanies chemical fixation of phytoplankton. Chemical fixation is used to stabilize cell structure for microscopic identification and to aid retention of ^{14}C activity during sample processing. Losses of particulate activity after fixation ranged from 0% to 86% (i.e., retention ranged from 14% to 100%) with a variety of fixatives including glutaraldehyde, formaldehyde, osmium tetroxide, potassium iodide, and potassium permanganate (Silver and Davoll, 1978). Differential losses are expected because of selective binding of the fixatives to different classes of macromolecules. For example, aldehydes are known to crosslink proteins, whereas osmium links lipids.

Freeze fixation has been recommended as an alternative to chemical fixation for the preparation of microalgal samples for autoradiography (Pearl, 1984). Rapid freezing in liquid nitrogen of phytoplankton concentrated on a membrane filter followed by freeze-drying consistently provided almost complete retention of isotope and minimized cell shrinkage and disruption, clearly outperforming chemical fixatives (Pearl, 1984). This provides the possibility of quantitative interspecific comparisons of photosynthesis rates in natural microalgal assemblages. Good recovery of ^{14}C may even be obtained in unfixed samples (Pearl and Stull, 1979) transferred to gelatin, as demonstrated for the chrococcoid cyanobacteria *Synechococcus* sp. (Iturriaga and Marra, 1988).

Autoradiographic techniques have enabled field measurements of the physiological responses of individuals to be integrated into investigations of changes in species abundance in phytoplankton communities. Despite the potential inefficiencies in retention and quantification of organic ^{14}C in single cells, autoradiographic investigations can provide information on the variability of cell-specific and cell-carbon-specific photosynthesis rates. Stull et al. (1973) estimated carbon-specific photosynthesis rates, which ranged from 0 to 0.58 h^{-1}, for lake phytoplankton. Corresponding doubling times ranged from 1.2 hr to 210 d, with 12 of 21 species having doubling times of less than 24 hr. Perhaps the most thorough investigation of phytoplankton dynamics that employed autoradiography was Knoechel and Kalff's (1978) study of productivity and succession of five diatom species during the spring and summer in Lac Hertel, Quebec. They concluded that growth rate was largely determined by light-saturated photosynthesis, which, in turn, responded to nutrient supply. However, net population increases were often less than the photoplankton growth rate primarily because of losses from sinking out of the photic zone, and sinking was, in turn, apparently related to senescence.

Isolation of Individual Cells

Rivkin and Seliger (1981) introduced scintillation counting of single phytoplankton cells as a technique for determining species-specific carbon assimilation rates. Previously, the same species-specific resolution in natural samples had only been attained for nearly monospecific blooms or for colonial microalgae such as tufts of *Oscillatoria*. Following a standard ^{14}C incubation, the large cells are concentrated on a 10 or 20 μm mesh screen, washed with nonlabeled medium, and finally suspended in a small volume of nonlabeled medium. Single cells are washed and transferred to a scintillation vial, then dried and hydrolyzed to remove inorganic ^{14}C. Activity is determined during long (up to 2 hr) counting times in a scintillation counter, with many samples having activities only a few counts per minute above background. The approach is applicable only to cells with diameters greater than approximately 10 μm because of limitations in the ability to manipulate smaller sizes and the low amount of ^{14}C that can be accumulated by small cells. The species most often examined include large robust dinoflagellates such as *Ceratium tripos*, *C. macrocystis*, and *C. fusus*, and large diatoms such as *Ditylum brightwellii* and *Coscinodiscus* spp. Taguchi and Laws (1985) also showed that the technique could be applied to fragile species such as the oceanic silicoflagellate *Dictyocha perlaevis*. The approach has been used extensively by Rivkin et al. (1982, 1984) and occasionally by other investigators. The major shortcoming of the technique is that only a small proportion of the total primary productivity may be due to the phytoplankton that can be examined using the single-cell isolation technique.

Perhaps the most interesting application of single-cell isolation is the ability to quantify the variability of cell-specific photosynthesis rates. Rivkin and Seliger (1981) found little intraspecific variation in ^{14}C uptake within samples with coefficients of variation for cell-specific ^{14}C-uptake rates ranging from 8 to 21% for laboratory cultures. Little variability was observed in cell chlorophyll content (c.v. = 15 to 23%) and photosynthesis rates (c.v. = 15 to 23%) among populations of the marine dinoflagellates *Pyrocystis noctiluca* and *P. fusiformis* from the surface mixed layer of the Sargasso Sea and Gulf Stream (Rivkin et al., 1982; 1984). In a hydrodynamically more variable environment, Boulding and Platt (1984) found much larger variations in cell-specific carbon assimilation by the dinoflagellate *C. tripos*, with coefficients of variation ranging from 29 to 73% in field samples. This higher variability may be a consequence of mixing populations that had previously adapted (phenotypically) to different environments within the water column.

Variability in photosynthesis rates of single cells may arise from phenotypic responses to differing preconditioning environmental variables and/or genetic differences between subpopulations or subspecies of morphologically identical form. Genetic variability appears to be the lesser of the two influences. For example, clonal variability in growth rates of morphologically identical cells isolated from a single water mass is typified by a coefficient of variation of less than 15% (Brand, 1985) a value remarkably similar to that found for cell-specific ^{14}C-uptake rates of pure cultures (Rivkin and Seliger, 1981). In contrast, phenotypic responses to variations in irradiance or nutrient availability may account for over an order of magnitude variation in average cell-specific photosynthesis rates (see Chapter 7). This phenotypic variability in photosynthesis rates has been demonstrated for natural populations of *Pyrocystis noctiluca* in which samples from the base of the euphotic zone were characterized by light-saturated photosynthesis rates, only 10% of those found for populations from the overlying mixed layer (Rivkin et al., 1982).

Recently, the same single-cell resolution has been obtained for cell division rates from experiments with tritiated thymidine incorporation (Rivkin and Voytek, 1986), allowing comparison of growth and photosynthesis rates in the same samples. Growth has been found to be less sensitive to irradiance than photosynthesis (Rivkin et al., 1984; Rivkin and Voytek, 1986) indicating that temporary uncoupling of photosynthesis and growth occurs in response to changes in the environmental conditions.

Tritium as a Tracer

McKinley and Wetzel (1977) suggested that neither carbon nor oxygen exchange would be suitable to determine primary productivity in oligotrophic waters with an exceptionally low buffering capacity. Under these conditions, the changes in O_2 concentration might be too small to measure and addition of ^{14}C-bicarbonate might alter the pH and significantly enrich the sample with inorganic carbon. McKinley and Wetzel (1977) proposed using tritiated water to measure primary productivity under these conditions. Unfortunately, this method suffers from a number of drawbacks, including high isotopic discrimination against [3]H$-$H$_2$O, high uptake of ^{3}H in dark bottles due to exchange of water of hydration, and large isotope enrichment necessitating the use of high specific activities in order to overcome dilution by unlabeled water. The isotope discrimination factor is 2.14 for [3]H$-$H$_2$O (Weinberger and Porter, 1953), compared to 1.06 for ^{14}C-bicarbonate. Dark-bottle ^{3}H accumulation was on average 74% of light-bottle accumulation in experiments reported by McKinley and Wetzel (1977). High specific activities (70 μCi·cm^{-3}) were used in experiments by McKinley and Wetzel (1977) in comparison to much lower activities of < 0.1 μCi·cm^{-3} typically employed in ^{14}C uptake experiments. One additional problem is determining the ratio of H$_2$O to CO$_2$ to O$_2$ used to compare results of ^{3}H incorporation experiments with photosynthesis rates obtained by other methods. McKinley and Wetzel (1977) found good agreement between [3]H$-$H$_2$O and ^{14}C-bicarbonate techniques, when an assimilation ratio of 2H:C was assumed. However, it is more likely that a ratio of 1.5–1.7H:C characterizes algae with a biochemical makeup that is more reduced than carbohydrate (Myers, 1980). Finally, it should be noted that there have been improvements in the precision of O$_2$ analyses since 1977 (Chapter 1), which may alleviate the need envisaged by McKinley and Wetzel (1977) for the use of [3]H$-$H$_2$O for measuring photosynthesis in oligotrophic, soft waters.

Stable Isotopes as Tracers

The radioactive isotope of carbon has long been the preferred tracer in photosynthesis experiments because of its availability, long half-life, and the high sensitivity of methods that employ [14]C. However, there are some circumstances in which it may be preferable to employ the stable isotope ^{13}C as a substitute for ^{14}C in tracer experiments. Carbon-13 can be used in conjunction with the stable oxygen isotope [18]O to simultaneously examine both CO$_2$ and O$_2$ production and consumption (Weis and Brown, 1959; Holmes et al., 1989) or with the nitrogen isotope ^{15}N to simultaneously measure carbon and nitrogen dynamics (Slawyk et al., 1988). Carbon-13 may become a prefered isotope for use in coastal and freshwater as increasingly stringent controls on the use of radioactive isotopes in the environment limit the locations in which ^{14}C uptake studies can be undertaken.

For continuous measurements of isotopic composition, a mass spectrometer is connected to a gas stream sampled from a small (cm^3 volume) experimental chamber via a gas-permeable membrane (Radmer and Ollinger, 1980). The gas stream is bombarded by an electron beam in the mass spectrometer, and the results are reported as a mass spectrum obtained on the basis of differences in the mass to charge (m/e) ratio of the ions produced. Most ions are singly charged, thus a mass spectrum would allow identification of $^{16}O_2$ ($m/e=32$), $^{18}O_2$ ($m/e=36$), $^{12}C^{16}O_2$ ($m/e=44$), and $^{13}C^{16}O_2$ ($m/e=45$). Determination of ^{13}C in plant material requires that the sample be combusted prior to introduction into the mass spectrometer. Dissolved oxygen can be extracted from aqueous samples for measurement of isotope exchange in dilute cultures or natural phytoplankton samples incubated in glass bottles (Bender et al., 1987; Kana, 1990).

Particulate Organic ^{13}C Production

The basic method for measuring photosynthesis by incorporation of $^{13}CO_2$ into particulate matter is similar to that described for the ^{14}C method. However, determination of the rate of photosynthesis by the ^{13}C method requires measurements of atom percent ^{13}C of the particulate matter at both the start and end of the experiment, and atom percent ^{13}C of the inorganic carbon and particulate organic carbon concentration at the start of the experiment. The carbon-specific assimilation rate is calculated from changes in atom percent ^{13}C as follows:

$$\rho = (1/C)dC/dt = (a_{t2} - a_{t1})/[(a_i - a_{t1})(t_2 - t_1)] \qquad (2.16)$$

where ρ is the rate constant for particulate carbon accumulation obtained from the isotope measurement, C is the particulate organic carbon concentration, a_{t1} and a_{t2} are the atom percent ^{13}C at the start (t_1) and end (t_2) of the experiment, and a_i is the atom percent ^{13}C of the inorganic carbon (Hama et al., 1983). To calculate the photosynthesis rate per unit volume of suspending solution requires the measurement of particulate organic carbon concentration:

$$(1/V)dC/dt = fC\rho \qquad (2.17)$$

where f is an isotope discrimination factor of 1.025 (Hama et al., 1983). Simultaneous measurement of particulate carbon concentration and atom percent ^{13}C is desirable to match the correct values of concentration and uptake rates and to reduce sample processing time. A combined automated particulate carbon and ^{13}C analysis employing a mass spectrometer in line with an automated elemental analyzer has been described by Preston and Owens (1985). New instrumentation allows the simultaneous examination of particulate organic carbon and nitrogen, together with ^{13}C and ^{15}N (Otsuki et al., 1983; Slawyk et al., 1988).

Measurement of photosynthesis by ^{13}C accumulation differs in several significant respects from those based on ^{14}C uptake. First, the stable carbon isotope ^{13}C is present at a much higher natural abundance ($^{13}C/^{12}C = 0.0011$) than its radioactive counterpart, necessitating correcting for the initial ^{13}C in a sample, whereas such a correction is unnecessary with the ^{14}C technique. Second, a large ^{13}C enrichment, which amounts to 5 to 15% of the original dissolved inorganic carbon is required because of the high natural abundance and limitations in the sensitivity of the detectors for measuring the $^{13}C/^{12}C$ ratio. In contrast, ^{14}C is usually added in truly trace amounts.

Hama et al. (1983) demonstrated that estimates of carbon assimilation from ^{13}C and ^{14}C methods gave excellent agreement for marine phytoplankton. However, good agreement is not always obtained in comparisons of ^{14}C and ^{13}C methods (Sakamoto et al., 1984; Slawyk et al., 1984). Methodological problems, such as nutrient enrichment of ^{14}C stock solutions stored in soft glass vials, systematic errors in POC measurement (Slawyk et al., 1984), or differences in the calculations used by different investigators to correct for dark accumulation (Sakamota et al., 1984), may account for these discrepancies. These problems highlight the need for a rigorous attention to detail when conducting either ^{14}C or ^{13}C experiments.

Gas chromatography can be coupled to mass spectrometry to determine the rates of synthesis of individual amino acids and monosaccharides in ^{13}C productivity experiments (Hama et al., 1987, 1988; Hama, 1988). The rates of carbohydrate and protein synthesis can be obtained by summing over the contributions of all monosaccharides and all amino acids, providing more accurate measurements of macromolecular synthesis rates than are possible with biochemical fractionation protocols used in ^{14}C experiments. Hama (1988) collected subsamples for total carbon uptake and for determination of amino acid and monosaccharide synthesis. The rates of synthesis of eight monosaccharides were calculated from ^{13}C atom percent in samples separated by gas chromatography following hydrolysis in H_2SO_4 for 24 hr at 100°C and conversion to acetyl derivatives. Rates of synthesis of 12 amino acids were estimated following acid hydrolysis at 105°C for 24 hr, after conversion to their corresponding N-trifluoroacetyl-*n*-butyl esters. These experiments indicated a high rate of glucose synthesis during the day, which probably accumulated as an energy reserve polymer. The ratio of glucose to total monosaccharides varied from 0.4 to 0.9 and the ratio of carbohydrate to protein varied from 0.5 to 2.8 in an upwelling area off Japan (Hama, 1988). During the night, a decline was observed in ^{13}C atom percent of glucose and increases were observed in ^{13}C atom percent in other monosaccharides and amino acids, indicating carbon flow from glucose to these other molecules, including proteins and structural polysaccharides, at night. Several complications arise in the interpretation of ^{13}C-labeling rates of individual compounds during these experiments. First, turnover of amino acids and carbohydrates cannot be independently evaluated from measurements of ^{13}C atom percent. Second, unlike the chlorophyll-labeling technique, which is specific for plant production, the amino acid and monosaccharide pools will include contributions from detritus, bacteria, and zooplankton. These problems can be partially overcome by calculating absolute production rates (i.e., the product of the concentrations and specific synthesis rates).

^{18}O measurements

There is no radioactive oxygen isotope with a half life suitable for use in photosynthesis experiments. However, use of the stable oxygen isotope ^{18}O allows a direct evaluation of gross photosynthesis (Bender et al., 1987; Grande et al., 1989) or of respiration in illuminated algae (Radmer and Ollinger, 1980).

$$2H_2\,^{16}O \Rrightarrow 4H^+ + {}^{16}O_2 \qquad\qquad (2.18)$$

$$^{18}O_2 + 4H^+ \Rrightarrow 2H_2\,^{18}O \qquad\qquad (2.19)$$

Mass spectrometric determinations of changes in $^{18}O_2$ and $^{16}O_2$ concentrations when employed in conjunction with specific metabolic inhibitors under variable environmental conditions provide a means of examining the rates of mitochondrial respiration, photorespiration, and the Mehler reaction in illuminated plants.

Most experiments that employ ^{18}O take advantage of the reaction shown in Eq. 2.19. To simultaneously measure respiration and photosynthesis in illuminated plants requires differentiation of the O_2 produced by photosynthesis from O_2 consumed by respiratory and nonrespiratory reactions. Since the natural abundance of ^{18}O is only 0.2% and the natural abundance of ^{16}O in water is 98.8%, essentially all of the oxygen evolved in photosynthesis will be $^{16}O_2$ (99.6%) with a minor contribution of ^{16}O

(0.4%), and an even smaller contribution of $^{18}O_2$ (0.0004%) (Radmer and Ollinger, 1980). Ninety-nine-percent pure $^{18}O_2$ is available and can be added to a sample to allow a significant enrichment of the dissolved O_2 with $^{18}O_2$. Thus, respiration (U) can be estimated from the consumption of $^{18}O_2$, whereas a simultaneous measurement of photosynthesis (P) is obtained from the evolution of $^{16}O_2$, as follows:

$$U = (\delta[^{18}O_2]/\delta t - k[^{18}O_2])([^{18}O_2] + [^{16}O_2])/[^{18}O_2] \qquad (2.20)$$

$$P_g = (\delta[^{16}O_2]/\delta t - k[^{16}O_2]) + U\{[^{16}O_2]/([^{16}O_2] + [^{18}O_2])\} \qquad (2.21)$$

where $[^{18}O_2]$ and $[^{16}O_2]$ are the mean concentrations of $^{18}O_2$ and $^{16}O_2$ (μmol·dm^{-3}) over a time period δt, $\delta[^{18}O_2]$ and $\delta[^{16}O_2]$ are the changes in $[^{18}O_2]$ and $[^{16}O_2]$ over the time interval of duration δt, and k is the rate of oxygen decrease due to consumption by the mass spectrometer (Peltier and Thibault, 1985).

These calculations assume that the $^{18}O_2/^{16}O_2$ isotope ratio measured in the suspending medium is the same as the isotope ratio at the site of O_2 consumption. The possibility that photosynthetic production of $^{16}O_2$ results in an intracellular dilution of the $^{18}O_2/^{16}O_2$ ratio can be tested by measuring changes in the rate of $^{18}O_2$ consumption between dark and light treatments under varying O_2 partial pressures. An inhibition of $^{18}O_2$ consumption in the light, which becomes more pronounced at low O_2 concentrations, provides evidence for isotope dilution at the site of oxygenase activity. Brown (1953) and Brown and Webster (1953) favor this explanation for observations on *Chlorella* and *Anabaena* in which a decline in $^{18}O_2$ consumption in the light relative to the rate in the preceding dark period was only observed under conditions of very low oxygen concentrations.

Oxygen Consumption in the Light

In pioneering experiments, Brown (1953) used $^{18}O_2$ to measure oxygen consumption by *Chlorella pyrenoidosa* in the light and the dark. He found no difference in oxygen consumption rates and concluded that mitochondrial respiration was neither stimulated nor inhibited by light. Although subsequent results appeared to conflict with these findings (see Graham, 1980), similar responses have been observed in other microalgae, including *Chlamydomonas reinhardtii* (Peltier and Thibault, 1985), and the diatom *Thalassiosira weisflogii* (Weger et al., 1989). Based on the responses to the inhibitors DCMU, KCN, and SHAM,* Bate et al. (1988) concluded that

---

* Dichlorophenyl dimethylurea (DCMU) inhibits photosynthetic oxygen evolution, potassium cyanide (KCN) inhibits energy conserving (cytochrome oxidase mediated) mitochondrial oxygen consumption, and salicylhydroxamic acid (SHAM) inhibits operation of the nonenergy-conserving (alternative) path of mitochondrial consumption.

Table 2.5. Ratio of O_2 consumption in the light (U) to O_2 consumption in darkness (r_d).

| Species[*] | U/r_d | Experimental Conditions | |
		Irradiance	Inorganic Carbon Supply
Anacystis nidulans	0.5–3	—	—
Chlamydomonas reinhardtii	0.6–1.0	700 $\mu mol \cdot m^{-2} \cdot s^{-1}$	TCO_2 compensation point
	1.6–2.0	700 $\mu mol \cdot m^{-2} \cdot s^{-1}$	TCO_2 saturation
Chlamydomonas reinhardtii	1–10	0-600 $\mu mol \cdot m^{-2} \cdot s^{-1}$	192 μM TCO_2
	1–2	0-600 $\mu mol \cdot m^{-2} \cdot s^{-1}$	980 μM TCO_2
Chlamydomonas reinhardtii	1	—	10 $mol \cdot m^{-3}$ TCO_2
Chlorellapyrenoidosa	1	150 ftc	—
Chondrus crispus	2	—	27 μM free CO_2
Cystoseira mediterranea	1.7	250 $\mu mol \cdot m^{-2} \cdot s^{-1}$	27 μM free CO_2
Dunaliella tertiolecta	0.6–1.1	1000 $\mu mol \cdot m^{-2} \cdot s^{-1}$	200 $\mu M TCO_2$
Enteromorpha compressa	1.3	250 $\mu mol \cdot m^{-2} \cdot s^{-1}$	27 μM free CO_2
Hydrodictyon africanum	2.8	—	10 $\mu mol \cdot m^{-3} TCO_2$
Hypenea musciformis	1.3	—	27 μM free CO_2
Ochromonas malhamensis	1–2	10–150 ftc	2% CO_2 in gasphase
Selenastrum minutum	1	—	1–2 $mol \cdot m^{-3}$
Thalassiosira weisflogii	1.7	150 $\mu mol \cdot m^{-2} \cdot s^{-1}$	2 $mol \cdot m^{-3} TCO_2$
	1.2	150 $\mu mol \cdot m^{-2} \cdot s^{-1}$	TCO_2 compensation point

[*]Sources of information are as follows: (1) Bate et al. (1988); (2) Brechignac et al. (1987); (3) Brown (1953); (4) Glidewell and Raven (1975); (5) Peltier and Thibault (1985); (6) Süeltemyer et al. (1986); (7) Weger and Turpin (1989); (8) Weger et al. (1989); (9) Weis and Brown (1959); (10) Süeltemyer et al. (1986); (11) Hoch et al. (1963).

mitochondrial respiration proceeded at a slightly lower rate in the light than in the dark. At reduced O_2 concentrations, an apparent light-dependent inhibition of mitochondrial respiration appears to be due to problems associated with intracellular isotope dilution (Brown and Webster, 1953). Table 2.5 provides a summary of ratios of O_2 consumption in the light to O_2 consumption in the dark for a number of micro- and macroalgae. Typically, the rate of O_2 consumption in illuminated cells (U) equals or exceeds the rate of dark respiration (r_d), although the ratio U/r_d can be extremely variable, ranging from 10 to 0.5 (Table 2.5).

Weger and Turpin (1989) showed enhancement of mitochondrial respiration following addition of ammonium or nitrate to nitrogen-limited

Table 2.6. Rates of O_2 consumption (μmol $O_2 \cdot mg^{-1}$chla$\cdot h^{-1}$) in darkened and illuminated *Selenastrum minutum* in the presence of various inorganic nitrogen sources.

| Treatment | Control | O_2 consumption rate | | |
		NH_4^+	NO_2^-	NO_3^-
Dark	100	220	130	110
80 μmol$\cdot m^{-2} \cdot s^{-1}$	110	250	150	130
230 μmol$\cdot m^{-2} \cdot s^{-1}$	110	200	150	110

Based on Weger and Turpin (1989).

chemostat cultures of freshwater microalgae. Addition of inorganic nitrogen stimulated amino acid synthesis, which in turn stimulated Krebs cycle activity and mitochondrial respiration at the expense of storage carbohydrates. The magnitude of this enhanced oxygen consumption was dependent on the form of nitrogen supplied but was largely independent of irradiance (Table 2.6). Oxygen consumption was stimulated to a lesser extent when NO_3^- was supplied to a nitrogen-deficient culture than when NH_4^+ was supplied (Weger and Turpin, 1989). Nitrate reduction could occur at the expense of reducing equivalents produced by the Krebs cycle, with NO_3^- substituting for O_2 as the terminal electron acceptor. Weger and Turpin (1989) ascribed essentially all of the oxygen consumption in these experiments with freshwater microalgae under saturating TCO_2 availability to mitochondrial respiration, with no evidence for either the Mehler reaction or photorespiration.

Oxygen consumption proceeds at rates that greatly exceed mitochondrial respiration in illuminated cells under some environmental conditions (Brown and Webster, 1953; Weis and Brown, 1959; Glidewell and Raven, 1975). The rate of O_2 consumption in illuminated cells can approach the rate of net O_2 evolution, which implies that gross photosynthesis may occur at twice the rate of net O_2 evolution (Kana, 1990).

Brown and Webster (1953) observed enhanced O_2 consumption in *Anabaena* when experiments were conducted under a combination of high O_2 concentrations and high irradiance. They suggested that this could have been associated with the Mehler reaction, but a subsequent investigation by Brown and Good (1955) found no evidence for photochemical oxygen reduction *in vivo* (Brown and Good, 1955). Other investigators (Glidewell and Raven, 1975; Süeltemeyer et al., 1986; Brechignac and Furbank, 1987; Brechignac et al., 1987), also using mass spectrometry, have found enhanced O_2 consumption in the light. Using inhibitors, these workers have shown that the enhanced O_2 consumption in the light can be attributed to the Mehler reaction at high CO_2 concentrations, and the Mehler reaction plus photorespiration at limiting CO_2 concentrations.

Table 2.7.   Rates of photosynthesis determined by various gas exchange techniques.

	Oxygen Exchange			
	Net O_2 Evolution	Dark O_2 Consumption	Gross O_2 Photosynthesis	Gross ^{18}O Evolution
Gymnodinium and *Pedinella* spp.	28.0 ± 2.3	7.4 ± 0.1	35.4 ± 2.4	43.5 ± 1.2
Thalassiosira pseudonona	63.7 ± 1.6	3.0 ± 0.1	66.7 ± 1.6	112.2 ± 12.1

	Carbon Exchange			
	Light CO_2 Consumption	Dark CO_2 Evolution	Gross CO_2 Photosynthesis*	Total ^{14}C Uptake
Gymnodinium and *Pedinella* spp.	28.4 ± 1.6	4.8 ± 1.0	33.2 ± 2.6	27.8 ± 5.6
Thalassiosira pseudonona	60.4 ± 1.6	0.4 ± 0.4	60.8 ± 2.0	85.2 ± 2.9

	Photosynthetic and Respiratory Quotients			
	Net Photosynthesis	Dark Respiration	Gross Photosynthesis*	$^{18}O/^{14}C$
Gymnodinium and *Pedinella* spp.	0.99	1.54	1.06	1.56
Thalassiosira pseudonona	1.05	7.5	1.10	1.32

After Bender et al. (1987).
*Gross photosynthesis is operationally defined as the sum of net photosynthesis and dark respiration rates.

An extensive comparison of different techniques for measuring algal gas exchange has recently been provided by Grande et al. (1989). In these experiments, the gross photosynthesis rate calculated as the sum of net oxygen evolution in the light plus oxygen consumption in darkness typically equalled the true rate of gross oxygen evolution determined using an ^{18}O technique (Fig. 2.4). Light respiration rates were calculated as the difference between gross ^{18}O photosynthesis and net oxygen evolution in the light. On average, light respiration exceeded dark respiration by 1.65 times (logarithmic mean), but there was considerable variability about this mean with the ratio of light-to-dark respiration ranging from 0.14 to 13. This large variability may be an inherent problem in attempting to compare small differences between two large values obtained for gross ^{18}O photosynthesis and net O_2 evolution with the much smaller values obtained for dark respiration rates. The elevated rates of respiration in

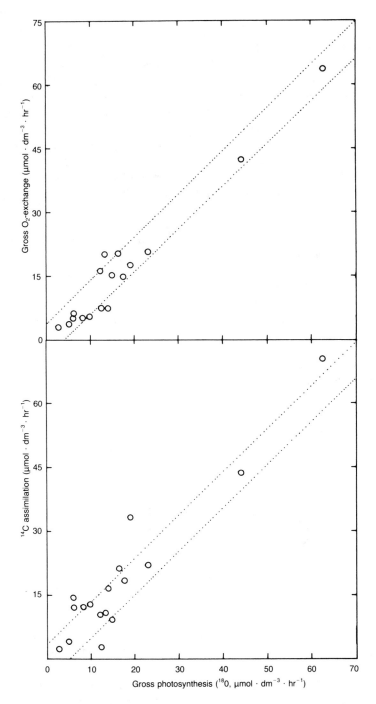

Fig. 2.4. Relationship between gross ^{18}O-exchange and light + dark O_2-exchange (a), or ^{14}C assimilation (b). Dotted lines refer to SD = ± 2. Data calculated from Grande et al. (1989).

the light were attributed to the Mehler reaction and/or photorespiration. However, no explanation was given for those observations in which dark respiration exceeded light respiration (Grande et al., 1989). Similarly, Kana (1990) has recently documented that oxygen consumption in the light may range from 0.75 to 10 times the rate of dark oxygen consumption in natural phytoplankton samples.

The potential magnitude or importance of light-dependent O_2 consumption processes in nature is illustrated in the observations of Bender et al. (1987) who found that the "true" rate of gross oxygen evolution obtained from the $H_2{}^{18}O$ technique exceeded the apparent rate of gross oxygen evolution calculated from net gas exchange (Table 2.7). Specifically, gross photosynthesis obtained from ^{18}O measurements was up to twice as high as gross photosynthesis calculated as the sum of net O_2 evolution in the light and consumption in darkness. The rate of oxygen consumption in the light was found to be approximately 20 times greater than O_2 consumption in darkness in a mixed assemblage dominated by the diatom *Thalassiosira pseudonana*, and approximately four times greater in an assemblage dominated by the flagellates *Gymnodinium* and *Pedinella*.

Isotope Discrimination and Natural Abundance

Use of the stable carbon isotope ^{13}C in studies of plant photosynthesis is not limited to tracer experiments as described previously. Measurements of natural abundance have aided in the evaluation of the physical and chemical processes that limit inorganic carbon assimilation and in the elucidation of different pathways of CO_2 fixation (O'Leary, 1988; Farquhar et al., 1989).

The natural abundance of ^{13}C in plant organic matter is expressed as the molar abundance ratio $R = {}^{13}C/{}^{12}C$ obtained for combusted plant material using a mass spectrometer. For natural materials $R \approx 0.00112$ with variation limited to the 5th decimal place, and for convenience results are reported as $\delta^{13}C$ which is defined as

$$\delta^{13}C = R_{sample}/R_{standard} \qquad (2.22)$$

where $\delta^{13}C$ has units of parts per thousand (‰) and the standard is limestone from the PeeDee formation in South Carolina (Farquhar et al., 1989).

$\delta^{13}C$ is a useful measurement in studies of photosynthetic carbon acquisition because carboxylation of CO_2 by RUBISCO discriminates markedly against the heavier isotope, whereas other carboxylations and transport processes are characterized by a very low discrimination against ^{13}C. For example, C_3 land plants are characterized by a $\delta^{13}C$ of about $-28‰$ whereas C_4 plants have a value of approximately $-14‰$. In general all plants contain less ^{13}C, and thus have more negative $\delta^{13}C$ values, than the medium in which they are suspended because of discrimination against the heavier isotope.

The changes in $\delta^{13}C$ that occur during a biochemical transformation or physical transport are expressed as $\Delta^{13}C$ for the transformation of compound A into compound B.

$$\Delta^{13}C = [\delta^{13}C(A) - \delta^{13}C(B)]/[1 + \delta^{13}C(B)] \qquad (2.23)$$

where Δ also has units of ‰. In contrast to the natural abundance (δ), the discrimination (Δ) is independent of the isotope composition of the standard used for the measurement (Farquhar et al., 1989). In addition, Δ is independent of source isotope composition and thus directly reflects biological and physical/chemical processes of interest. However, the use of Δ does require a knowledge of the inorganic carbon source (CO_2 or HCO_3^-). At low temperatures CO_2 and HCO_3^- can have $\delta^{13}C$ values that differ by more than 10‰.

The $\delta^{13}C$ of micro- and macroalgae varies from $-6‰$ to $-35‰$ (Smith and Walker, 1980; Descolas-Gros and Fontugne, 1990), which originally suggested to some investigators that these plants possessed both C_3 and C_4 photosynthetic pathways. However, it now seems unlikely that the algae have a C_4 photosynthetic pathway (Smith and Walker, 1980; Farquhar et al., 1989; Descolas-Gros and Fortugne, 1990). The general low (less negative) values implied for $\delta^{13}C$ in aquatic plants when compared to the expected discrimination by the C_3 pathway have been attributed to the importance of diffusive processes in limiting inorganic carbon acquisition, or to the operation of an inorganic carbon concentrating mechanism (Smith and Walker, 1980; Sharkey and Berry, 1985; Descolas-Gros and Fortugne, 1990). Diffusion limitation would result in a low isotopic discrimination, because essentially all of the CO_2 that reached RUBISCO would be consumed. Many algae appear to possess a mechanism that raises the intracellular concentration of CO_2, and, in these plants, fractionation during membrane transport may account for the isotopic discrimination, again because most of the CO_2 that reaches RUBISCO will be fixed (Berry, 1988). In *Chlamydomonas reinhardtii* the effect of active inorganic carbon accumulation is to reduce $\Delta^{13}C$ from approximately 28‰ in cells grown in 5% CO_2 to 4‰ in cells acclimated to air levels of CO_2 (Sharkey and Berry, 1985).

The correlation between $\delta^{13}C$ and the ratio of activities of carboxylation by RUBISCO and β-carboxylases in marine microalgae was earlier attributed to variations in the rates of RUBISCO and β-carboxylations (Descol-as-Gros and Fontugne, 1985, 1990). This interpretation may need to be revised in view of the often low rates of β-carboxylation in the algae, and the roles of diffusive limitation and/or inorganic carbon concentrating mechanism on $\Delta^{13}C$. Perhaps the relation between $\delta^{13}C$ and the RUBIS-CO:PEP carboxylation ratio as microalgal cultures progress from exponential to stationary phase is due to a methodologic artifact such as a reduction in the total inorganic carbon concentration in the suspending medium or to changes in the bulk biochemical composition of the algae. In this regard, between and within plant variability in ^{13}C, natural abundances of up to 8‰ have been documented in the marine macroalgae *Laminaria longicruris* (Stephenson et al., 1984) and *Ecklonia radiata* (Fenton and Ritz, 1989), and this variability has been attributed to changes in the relative amounts of different macromolecular pools (lipid, carbohydrate, and protein) that are characterized by different isotopic compositions. Nevertheless, low $\delta^{13}C$ values in species from the genus *Prorocentrum* are always associated with high rates of PEP carboxylation (Descolas-Gros and Fontugne, 1990), and this requires further investigation.

Chapter 3

Fluorescence Techniques

Fluorescence measurements have been used extensively in phycological research for estimating chlorophyll concentration in situ using techniques developed by Lorenzen (1966). Despite the fact that fluorescence from chlorophyll a in vivo is extremely low ($\approx 3\%$, Latimer, Bannister & Rabinowitch, 1956) it provides a measure of pigment content with a sensitivity considerably greater than that of the limiting sensitivity of absorption measurements. Fluorescence measurements also provide information about competing mechanisms involved in the decay of excitation energy during photochemical reactions and, in recent years, have been used as an adjunct to gas exchange determinations, providing additional information on mechanisms underlying variations in carbon assimilation. Fluorescence techniques have an advantage over many conventional procedures in that they can provide a sensitive, nonintrusive probe of the photosynthetic apparatus. Considerable evidence now exists to indicate that alterations in fluorescence kinetics are closely correlated with changes in carbon metabolism and our present understanding of functional aspects of the photosynthetic apparatus have been largely brought about through the increased application of fluorometric techniques to photosynthesizing cells. For this reason, this chapter considers those applications of fluorescence that are relevant to an understanding of gas exchange responses. It is often customary to compare fluorescence and photosynthesis measurements made over a time scale of minutes and to ignore fast (micro- to millisecond) fluorescence kinetics. However, this approach neglects the fact that any one of a number of reactions varying on time scales of fractions of a second to minutes could have important effects on the rate of net CO_2 fixation. For this reason, we have included information on rapid fluorescence transients. Because the interpretation of fluorescence emission characteristics depends on a knowledge of the molecular processes underlying light emission, these are also discussed briefly. Whether fluorescence measurements in themselves are sufficient to characterize photosynthetic responses is currently uncertain, and fluorescence measurements are probably best used in combination with gas exchange determinations. Recent reviews and information on the theory and measurement of fluorescence parameters can be found in Papageorgiou (1975), Bolhar-Nordenkampf et al. (1989), Geacintov and Breton (1987), and in the book *Light Emission by Plants and Bacteria* edited by Govindjee et al. (1986). For details of current instrumentation, the reader should consult Mauzerall (1980), Schreiber (1983), Schreiber and Schliwa (1987), Ogren and Baker (1985), Schreiber, et al. (1989), and Bolhar-Nordenkampf et al. (1989). Details of the application of fluorescence measurements to ecological research are given in Schreiber and Bilger (1987), Bolhar-Nordenkampf et al. (1989), Krause and Somersalo (1989), Schafer and Björkman (1989) and Kiefer et al. (1989). Unfortunately, most of the more recent studies of fluorescence have been made on terrestrial vascular plants and further information on algae is required.

Deexcitation Processes and the Origins of Chlorophyll Fluorescence

To accurately interpret the functional significance of variations in fluorescence emission requires a detailed understanding of the mechanism(s) underlying light emission by plants in vivo. Fluorescence generally occurs from the lowest energy level of the excited singlet state and has a peak emission at longer wavelengths than the source inducing fluorescence. In addition to fluorescence, excitation energy can be dissipated by phosphorescence or thermal decay as shown in Fig. 3.1. Currently, the formation of triplet states (reverse spin orientation of the excited molecule) and consequent emission of light by phosphorescence is thought to be unlikely in vivo (but see Sane et al., 1974). However, modifications in the localized environmental conditions surrounding macromolecular structures can greatly increase light emission by phosphorescence (Vanderkooi and Berger, 1989) and could result in difficulties in the interpretation of fluorescence kinetics. Phosphorescence can be distinguished from fluorescence by its greater lifetime and by a longer wavelength emission. Loss of excitation energy within an excited state or via overlapping ground and excited states can occur through radiationless (dissipative) decay as thermal energy (Fig. 3.1.) and this is now thought to have a considerable influence on variations in photosynthetic performance and fluorescence emission.

At room temperature, most of the fluorescence is due to Photosystem

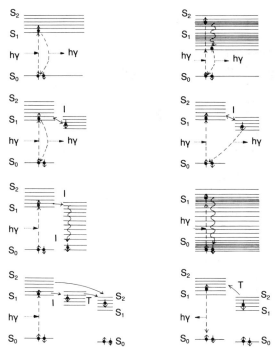

Fig. 3.1. Possible routes contributing to the decay of excitation energy, including photochemical quenching. Molecular orbitals are denoted by horizontal lines with differences between lines representing variations in energy levels within an excited state and the direction of electron spins indicated by arrows. Absorption/emission of a photon ($E = h\nu$) is indicated by the dotted lines connecting the ground state (S_0) with the excited states (S_1, S_2), where S_1 and S_2 represent the lowest and highest energy levels within an excited state, respectively. I denotes intersystem crossing over to the triplet state resulting in electron spin reversal and T denotes electron transfer. (a) Prompt fluorescence emission. (b) Radiationless decay followed by fluorescence emission. (c) Delayed fluorescence involving intersystem crossing and triplet formation. (d) Formation of the triplet state and phosphorescence emission. (e) Radiationless decay via the triplet state. (f) Radiationless decay via overlapping excited and ground states. (g) Photochemical charge separation via either singlet or triplet states. (h) Charge recombination after primary charge separation resulting in delayed fluorescence emission. Adapted in part from information in Clayton (1970).

2 (PS2) with a peak wavelength at $\lambda = 685$nm, although there is evidence for some Photosystem 1 (PS1) emission at longer wavelengths (Geacintov and Breton, 1987; Bolhar-Nordenkampf et al., 1989). In general, similar characteristics are found in all algal groups. The alga *Porphyra perforata* seems to be exceptional with regard to its room temperature emission peak, with a major band at 730 nm due to PS1 fluorescence (Fork et al., 1982). At the temperature of liquid nitrogen (77°K) significant fluorescence emission occurs from PS1 in all algae (see below). In the phycobilin-containing algae, significant emission can also be measured at $\lambda = 653$ nm

due to the presence of phycobilins, if the actinic light is absorbed by both chlorophyll a and phycobiliproteins (see Fork and Mohanty, 1986).

Fluorescence decay with lifetimes of the order of 2 ns is called *prompt* (variable) fluorescence, whereas light emission on time scales ranging from 25 ns to > 90 s is called *delayed* fluorescence, but both have the same emission spectrum. Prompt fluorescence is generally considered to be due to the emission of photons directly from the antennae system of PS2. However, there is still considerable uncertainty about the origins of prompt fluorescence (Hansson and Wydrzynski, 1990), which may also be due to a very rapid charge recombination reaction (Klimov and Krasnovskii, 1983), at the reaction center of PS2. Although delayed fluorescence could result from triplet formation, it is thought to be largely due to charge recombination of the primary photoproducts P^+_{680} and reduced phaeophytin (Pheō), when the secondary electron acceptor (Q_a) is fully reduced. The heterogeneous nature of the delayed fluorescence (Hipkins, 1978), may indicate, however, that more than one process can contribute to this type of light emission.

Using PS2 reaction centers with or without the primary electron acceptor Pheo$^-$, Barber et al. (1989) have shown that the major portion of the variable fluorescence is not due to a back reaction involving the radical pair P^+_{680} and Pheo$^-$ but is due to decay of the excited singlet state. This does not, however, entirely rule out a contribution from a recombination process (Barber et al., 1989). We can outline these various possibilities leading to fluorescence emission in the following way:

$$h\nu + DP^*_{680} \, \text{Pheo} \, Q_a \Rightarrow DP^+_{680} \, \text{Pheo}^- Q_a \tag{3.1}$$

$$DP^+_{680} \, \text{Pheo}^- \, Q_a \Rightarrow DP^+_{680} \, \text{Pheo} \, Q^-_a \tag{3.2}$$

$$D^+P_{680} \, \text{Pheo} \, Q^-_a + h\nu \Rightarrow D^+ \, P^*_{680} \, \text{Pheo} \, Q^-_a \tag{3.3}$$

$$D^+P^*_{680} \, \text{Pheo} \, Q^-_a \Rightarrow D^+ \, P_{680} \, \text{Pheo} \, Q^-_a + h\nu \tag{3.4}$$

$$D^+P_{680} \, \text{Pheo} \, Q^-_a \Rightarrow D^+P^+_{680} \, \text{Pheo}^- \, Q^-_a \tag{3.5}$$

$$D^+P^+_{680} \, \text{Pheo}^- \, Q^-_a \Rightarrow D^+P^*_{680} \, \text{Pheo} \, Q^-_a \tag{3.6}$$

$$D^+P^*_{80} \, \text{Pheo} \, Q^-_a \Rightarrow D^+P_{680} \, \text{Pheo} \, Q^-_a + h\nu \tag{3.7}$$

where $h\nu$ signifies a photon, D is the electron donor of PS2, and P^*_{680} is the excited state of the reaction center of PS2. Reaction 3.4 corresponds to prompt fluorescence, whereas reaction 3.7 results in delayed fluorescence. Reaction 3.4 will be in direct competition with other processes that can deactivate the excited state, whereas reaction 3.7 will depend on the rate of charge recombination between $P^*_{680}{}^1$ and Pheo$^-$ and will only occur when Q_a is fully reduced so that electron transfer from Pheo$^-$ is not possible.

Fluorescence Kinetics

Fluorescence is only one of a number of ways in which excitation energy can be released returning the excited molecule to the ground state. The efficiency with which fluorescence emission occurs can be quantified by the use of the photon yield of PS2 fluorescence (= fluorescence yield, ϕ_f)

$$\phi_f = F/\beta A \tag{3.8}$$

where F is the fluorescence emission, A is the amount of light absorbed, and β is the fraction of the absorbed light diverted to PS2.

Butler (1978) pioneered the use of rate constants to quantify the various processes that are involved in the deexcitation of the excited singlet state. The probability of a particular deexcitation process occurring is specified by a rate constant (k) with units of inverse time. The photon yield of PS2 fluorescence is given by the rate constant for fluorescence deexcitation divided by the rate constants of all the competing reactions. Equation (3.8) then becomes

$$\phi_f = k_f/\beta(k_f + k_d + k_t + k_{ph} + k_p \cdot P) \tag{3.9}$$

where the subscripts f, d, ph, and p refer to the rate constants for fluorescence, radiationless dissipation, energy transfer to PS1, phosphorescence, and photochemistry respectively. Phosphorescence is thought to be minimal in vivo and k_{ph} is generally ignored. The rate constant for photochemistry, k_p, is generally multiplied by the fraction of open PS2 reaction centers P. When $P = 1$ fluorescence emission is minimal (called F_0) as all of the reaction centers are open and when $P = 0$ fluorescence emission is maximal (called f_m) as all of the reaction centers are closed.

The lifetime of the decay of fluorescence can provide information on the type and origin of the fluorescence signal. The lifetime of a particular deexcitation process is simply the inverse of the rate constant for that process. If the only possible route for decay is through fluorescence then,

$$\tau_0 = 1/k_f \tag{3.10}$$

where τ_0 is the natural or intrinsic lifetime of the excited state. However, in a complex system such as the photosynthetic apparatus, there are several possible routes for the decay of the excited state, each of which is determined by its own particular rate constant. The real or actual lifetime, τ_a, is determined by the rate constants of all the competing processes.

$$\tau_a = 1/\beta(k_f + k_d + k_t + k_p P) \tag{3.11}$$

From Eqs. 3.9 and 3.11, $\phi_f = k_f \tau_a$ and from Eq. 3.10

$$\phi_f = \tau_a/\tau_0 \tag{3.12}$$

Therefore, the photon yield for fluorescence equals the actual lifetime divided by the intrinsic lifetime of the excited state.

The Kautsky Curve

On irradiating previously darkened photosynthetic tissue the fluorescence emission exhibits a number of characteristic transients (Fig. 3.2), which were first described by Kautsky and Hirsch (1931). In their original experiments Kautsky and Hirsch (1931) suggested that these changes were related to photosynthetic processes, although it was not until the experiments of Duysens and Sweers (1963) that the complementary relationship between fluorescence and photosynthesis gained increased acceptability. The observed changes in fluorescence emission are often extremely complex and will vary according to the length of the dark period used to ensure complete oxidation of the intersystem electron carriers, the physiological condition of the tissue, and the previous environmental conditions. Qualitatively, similar transients are found with all algal groups (Fork and Mohanty, 1986; Govindjee and Satoh, 1986), which are also similar to those found with terrestrial vascular plants (Briantis et al., 1986). Inflections in the Kautsky curve are often labeled OIDPMST (see below), although some authorities have chosen to subdivide the S and M transients, because of the presence of additional features on the fluorescence emission curve. Normally, the maximum fluorescence (F_m) is found at P, although with *Anacystis nidulans* the highest emission is found during the S–M transient (Mohanty and Govindjee, 1973). Oscillations associated with the M peaks have been correlated with the initiation of CO_2 assimilation (Walker et al., 1983; Ireland et al., 1984).

Changes in fluorescence emission during the Kautsky transients have been largely attributed to variations in the redox state of the secondary electron acceptor Q_a, although these may be modified by nonphotochemical processes and by physical factors related to variations in the absorp-

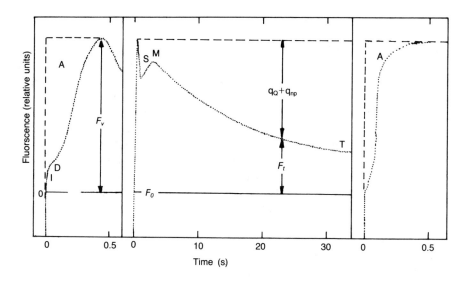

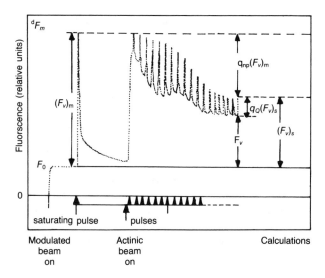

Fig. 3.2. Fluorescence emission characteristics (Kautsky curve) of previously darkened plant material. The overall changes in fluorescence emission are shown in panel (b), which illustrates the typical Kautsky transients. The fast fluorescence rise is shown in the absence (a) and presence (c) of the inhibitor DCMU, which prevents the reoxidation of Q_a. A is the complementary area above the initial fluorescence rise. Photochemical and nonphotochemical quenching components can be determined as in (d) using a modulated fluorometer and a train of saturating light pulses superimposed on the fluorescence emission produced by a continuous actinic source (see text for details). F_0 is the minimum level of fluorescence, F_m is the maximum level of fluorescence and F_t is the fluorescence emission at any point during the Kautsky transient. q_E and q_{NP} are the photochemical and nonphotochemical quenching components, respectively. In panel (d) $(F_v)_m$ is the maximum level of variable fluorescence and $(F_v)_s$ is the level of variable fluorescence during a saturating light pulse. Adapted and redrawn from the data of Bolhar-Nordenkampf et al. (1989).

tion cross section of PS2 (Mauzerall, 1972). Detailed descriptions of current interpretations of the fluorescent transients are given in Papergeorgiou (1975), Krause and Weis (1984), Govindjee and Satoh (1986), Fork and Mohanty (1984), and Geacintov and Breton (1987). The following is a summary of the main processes that are thought to underly the characteristic changes in fluorescence emission during induction.

Immediately after a dark-adapted sample is illuminated the fluorescence emission increases to an initial level (0) termed F_0. This initial fluorescence is thought to be due to emission from excited antenna chlorophyll prior to the migration of excitation energy to the reaction centers, including a significant contribution from PS1 antenna chlorophylls (see Geancintov and Breton, 1987). Experimentally, the resolution of the "true" F_0 will depend on the speed of the shutter opening (Bolhar-Nordenkempf et al., 1989) and on the length of the initial dark period, as this will determine the redox state of Q_a. F_0 should only be determined when all of the reaction centers are open and Q_a is fully oxidized. Obviously it is important to obtain an accurate value of F_0 as the calculated (variable) fluorescence ($F_v = F_m - F_0$) is used to determine a number of kinetic parameters. The F_0 emission will be related to photosynthesis, because it is dependent on factors that influence energy distribution between PS1 and PS2, such as state transitions, and on the structural integrity of the PS2 (or PS1) complex. For instance, the F_0 level can be reduced as a result of photoinhibitory or thermal damage (see Krause and Weis, 1984). Quenching of F_0 fluorescence may also occur under some conditions (Bilger and Schreiber, 1986; Horton and Hague, 1988; Hodges et al., 1989), so that it cannot be considered as a constant background level of emission.

The 0–I rise reflects the initial rapid reduction of Q_a by PS2 and the subsequent decline or plateau from I to D is associated with a temporary reoxidation of Q_a by PS1 (Krause and Weis, 1984). The 0–I rise precedes the I–D transient because of the finite time interval required for PS1 to oxidize Q_a because of the large plastoquinone pool. There then follows a marked increase in fluorescence emission (the D–P rise) caused by the subsequent reduction of all of the electron carriers between Q_a and the electron acceptors on the reducing side of PS1. Govindjee and Satoh (1986) suggested that electron transfer between ferredoxin and $NADP^+$ is restricted in the dark-induced state and that complete reduction of Q_a only occurs after the reduction of electron carriers on the acceptor side of PS1. Beyond this peak P, termed F_m, fluorescence declines. Reoxidation of PS1 acceptors and an increase in electron flow through PS1 is correlated with the initiation of carbon assimilation and the reoxidation of Q_a. However, the decline in fluorescence during this period is not linearly related to the oxidation state of Q_a (Lavorel et al., 1986), because it is due to both photochemical (reoxidation of Q_a) and nonphotochemical (unrelated to the redox state of Q_a) processes. At T, fluorescence emission reaches a stationary state corresponding to steady-state levels of CO_2 assimilation.

At high irradiances, the I–D dip is largely eliminated, and the fluorescence emission rises sharply to the peak fluorescence F_m (Schreiber, 1986). Recently three phases in the transition from F_0 to F_m have been identified using a combination of fluorescence measurements with P_{700} determinations (Schreiber et al., 1989). Under these conditions, it is unlikely that any Q_a is oxidized, and these three phases have been attributed to Q_b reduction, PQ reduction and PS1 reduction respectively. Once Q_a is fully reduced Pheo$^-$ may transfer electrons directly to PQ. Under these conditions, the main reason for the fluorescence increase is due to a reduction of both PQ and Q_b (Schreiber et al., 1989), indicating that at high irradiances other components of the electron transport chain, besides Q_a, can quench fluorescence.

As well as providing information on the redox state of Q_a, the fluorescence emission curve can also provide additional information on a number of photosynthetically related parameters. The generally sigmoid nature of the fluorescence rise is thought to reflect energy transfer between PS2 complexes as excitation energy is transferred from closed to open reaction centers. The complementary area above the fluorescence induction curve (Fig. 3.2) can also be used to estimate the acceptor pool size of PS2 (Murata et al. 1966; Schreiber et al., 1989), although this is critically dependent on an accurate measurement of maximal fluorescence (Bell and Hipkins, 1985) and F_0. Although the analysis of the complementary area is complicated (Lavorel et al., 1986; Geacintov and Breton, 1987), comparisons between experimental treatments can be made using the half time ($t_{0.5}$) of the rise from F_0 to F_m (Bolkar-Nordenkampf et al., 1989).

In the presence of diuron (DCMU), which blocks electron transfer from Q_a to Q_b, fluorescence emission is thought to be solely due to Q_a reduction and the complementary area can be used to measure the pool size of Q_a (Malkin and Kok, 1966). Using this information and the results of the complementary area above the fluorescence curve in the absence of DCMU, the pool size of PQ can also be determined by difference. Solving for the kinetics of the increase in the complementary area initially indicated two components that were attributed to two different forms of PS2, called PS2$_\alpha$ and PS2$_\beta$ (Melis, 1984). It has been proposed that the α centers are fully functional in electron transport, whereas the β centers are largely inactive (Melis and Anderson, 1983) and may be converted into α centers (Melis, 1985). However, these early analyses have been criticized because of the difficulty of accurately measuring the asymptotic level of F_m (Bell and Hipkins, 1985). Because of this problem, Hsu et al. (1989) derived F_m

from a mathematical analysis of the fluorescence kinetics. Their analysis indicated three phases with rate constants of 173, 4.45, and 1.1 s^{-1}, the slowest phase of which was not found in earlier analyses. Existing hypotheses regarding the occurrence and functional significance of the α and β centers may therefore require further examination. It is also not clear how the α and β centers are related to the two different redox forms of Q_a (Horton and Crose, 1979; Hansson and Wydrznaki 1990), although current evidence indicates that both forms of Q_a are associated with PS2$_\alpha$ (Thielin and Van Gorkom, 1981). Because other fluorescence techniques indicate a single population of PS2 (Mauzerall and Greenbaum, 1989), this matter has yet to be resolved.

Fluorescence Quenching Components

Reductions in fluorescence from a peak level where all of the reaction centers are temporarily closed to the terminal, or steady-state, level of fluorescence can occur via several routes. The extent to which each process reduces or "quenches" fluorescence is generally quantified using quenching coefficients. The quenching coefficient (q) can vary from 0 (peak fluorescence emission) to 1 (no fluorescence emission). Basically there are two ways fluorescence can be quenched; either by photochemically induced charge separation (i.e. transfer of an electron from Pheo$^-$ to Q_a), called photochemical quenching (q_Q), or by processes unrelated to photochemistry, a composite term called nonphotochemical quenching (q_{NP}). Nonphotochemical quenching has been resolved into (1) energy-dependent quenching (q_E), the major nonphotochemical quenching component, which is thought to be largely related to the magnitude of the transthylakoid ΔH^+ gradient (Krause et al., 1982) and/or zeaxanthin formation (Schafer and Bjorkman, 1989), (2) state transition-dependent quenching (q_T), which is correlated with protein phosphorylation of PS2 and alterations in the distribution of excitation energy, and (3) photoinhibition-dependent quenching (q_I), which occurs by an as yet unidentified mechanism (see Horton and Hague, 1988; Hodges et al., 1989) but may also be due to zeaxanthin formation (Demmig et al., 1987). Interestingly, the sum of all the quenching processes has been found to be constant over wide variations in irradiance, indicating that they may be actively involved in the regulation and control of the dissipation and utilization of excitation energy in response to the current requirements of the photosynthetic apparatus (Horton and Hague, 1988).

There is considerable interest in the factors that control the quenching of fluorescence during the decline from F_m to steady state. During induction, these kinetics occur on time scales that are correlated with the initiation of carbon metabolism. It is now largely accepted that q_Q and q_{NP} quenching underly the fluorescence decline, and there is increasing evidence to indicate that interactions between q_Q and q_{NP} are actively involved in the regulation of carbon metabolism (Weis and Berry, 1987; Horton and Hague, 1988; Holmes et al., 1989; Schafer and Björkman, 1989). Although many earlier analyses have related variations in q_Q and q_{NP} during fluorescence induction to steady-state measurements of carbon fixation, it is now possible with pulse-modulated systems to measure fluorescence kinetics and photosynthesis at the same time (Weis and Berry, 1987; Holmes et al., 1989; Schafer and Björkman, 1989). As yet, however, these techniques have not been widely used in phycological research, and much of the commercially available instrumentation was developed for ecophysiologic work on terrestrial vascular plants (see Schreiber and Bilger, 1987; Bolhar-Nordenkampf et al., 1989). Nevertheless, these techniques are directly applicable to studies of algal photosynthesis (see Holmes et al., 1989).

Resolution of Photochemical and Nonphotochemical Quenching

Earlier experiments, generally with isolated thylakoids or algal cells (Krause et al., 1982), used the relaxation (increase) in fluorescence on the addition of DCMU to estimate q_Q and q_{NP}. These kinetics can be resolved experimentally into two components, a rapid phase ($t_{0.5} = 1$ s) due to the rapid reduction of Q_a and a slower phase ($t_{0.5} = 5$–15 s) due to the relaxation of the transthylakoid proton gradient. However, the DCMU-addition technique has a limited application in many studies because of its slow rate of penetration into intact organelles, whole cells, or tissues. Indeed, some workers (Hodges and Barber, 1983) have related differences in fluorescence kinetics to variations in DCMU-induced reoxidation of Q_a. Furthermore, DCMU poisoned samples cannot be reused for subsequent analyses, and any technique associated with the use of inhibitors to examine *in vivo* responses should be confirmed using alternative procedures.

To overcome problems associated with the interpretation of DCMU-induced relaxation kinetics, Bradbury and Baker (1981) introduced a light-doubling method that relies on differences in the relaxation kinetics of q_Q (fast) and q_{NP} (slow) processes. Application of a short, high irradiance light pulse that fully reduces Q_a will saturate q_Q, but in theory should not produce any change in q_{NP}. Obviously, the duration of the pulse is critical to obtain an accurate measure of q_Q; if the pulse is too short, Q_a will not be fully reduced, if the pulse is too long, nonphotochemical quenching processes will be induced. In modern developments of this technique, modulated fluorometers are used, which overcome problems associated with the earlier requirement to determine fluorescence emission at a range of irradiances and greatly simplifies the analysis used by Bradbury and Baker (1981).

A continuous actinic beam can be used to produce steady-state rates of carbon assimilation and generate typical Kautsky kinetics. Following the fluorescence peak (F_m) a series of saturating light pulses are applied to fully reduce Q_a. The maximum fluorescence emission for each light pulse declines with time (Fig. 3.3), and the difference between the peak emission and F_m increases. Because each pulse should be intense enough to fully reduce Q_a, the remaining quenching must be due to nonphotochemical mechanisms. q_Q and q_{NP} can be calculated using the procedure described by Schreiber et al. (1986) (Fig. 3.2).

$$q_Q = (F_v)_s - F_v/(F_v)_s \qquad (3.13)$$

$$q_{NP} = (F_v)_m - (F_v)_s^*/(F_v)_m \qquad (3.14)$$

where $(F_v)_s$ is the saturating fluorescence during a light pulse and $(F_v)_m$ is the maximum variable fluorescence. Normalizing q_Q and q_{NP} to the maximum "available" variable fluorescence [either $(F_v)_s$ for q_Q or $(F_v)_m$ for q_{NP}] allows their relative contribution to the total quenching to be estimated.

Resolution of Nonphotochemical Quenching Components

Earlier results indicated that a proportion of the nonphotochemical fluorescence quenching could not be described solely to energy-dependent quenching (q_E) by the thylakoid proton gradient. Despite this, measurements of q_{NP} often provide good estimates of q_E (Schafer and Björkman, 1989), indicating that energy-dependent quenching is always a large proportion of q_{NP}, particularly at high irradiances (Horton and Hague, 1988). In a recent analysis, Horton and Hague (1988) have used additions of the inhibitors DCMU, NaF, and chloramphenicol during dark relaxation of fluorescence to estimate the contributions of protein-phosphorylation-dependent changes in excitation energy distribution (q_T) and photoinhibition (q_I), to q_{NP} using modulated fluorescence. On the addition of DCMU without background light but with the pulse-modulated source, an additional phase of relaxation was observed, which was not due to q_E, but was sensitive to NaF addition and ascribed to thylakoid protein phosphorylation, as NaF is an inhibitor of the thylakoid protein phosphatase enzyme. At low irradiances the quenching by q_T approached F_m, but at high irradiances there was a significant difference between the

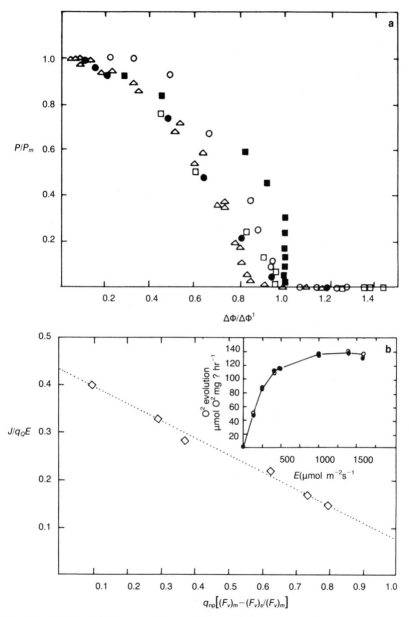

Fig. 3.3. Relationship between photosynthesis and fluorescence of some eucaryotic algae. (a) The relative photosynthetic capacity, P/P_m, is plotted against the relative fluorescence yield, $\triangle\o/\triangle\o'$, normalized to 1.0 at a $\triangle\o$ corresponding to a P/P_m of 0.01. Symbols refer to *Skeletonema costatum* (●), *Isochrysis galbana* (□), *Thalassiosira weisflogii* (△), *Dunaliella tertiolecta* (■), and *Chlorella vulgaris* (○). After Falkowski et al. (1986). (b) $J/q_Q.E$ for *Scenedesmus minutum* is plotted against the phototchemical quenching coefficient q_{NP} estimated from Eq. 3.21 Rates of steady-state electron flow were calculated from measurements of gross O_2 exchange, determined by mass spectrometry and q_{NP} was varied by changing the incident irradiance. Inset to (b) shows the predicted (○) and observed (●) rates of photosynthetic electron flow in response to variations in irradiance. After Holmes et al. (1989). For a full description and an explanation of the symbols see text.

measured fluorescence and F_m. This was attributed to photoinhibition-dependent irreversible quenching (q_I). If these components are normalized using the following formula (Horton and Hague, 1988), each can be expressed as a proportion of the total quenching ($q_Q + q$).

$$qX = \Delta F_x / (\Delta F_x + F'_x) \tag{3.15}$$

where X refers to q_Q, q_E, q_T, and q_I, ΔF_x is the amplitude of the fluorescence increase due to quenching process X and F'_x is the level of fluorescence reached after relaxation of the previous quencher minus F_0. This analysis may not be applicable to intact tissues because of uncertainties due to the use of inhibitors. Horton and Hague (1988) have used measurements of dark relaxation in the presence of saturating pulses, which would give an estimate of changes in the oxidation state of Q_a, in an attempt to eliminate these difficulties (see also Hodges et al., 1989). On the basis of different phases in the dark-relaxation curves, they could identify a fast phase ($t_{0.5} = 30$ s), an intermediate phase ($t_{0.5} = 8$ min) and a slower phase ($t_{0.5} = 30$ min), which were consistent with the inhibitor experiments on isolated protoplasts and related to q_E, q_T, and q_I quenching (Horton and Hague, 1988). Schafer and Björkman (1989) could only resolve the dark relaxation kinetics into ($t_{0.5} = 1$–2 min and 45–116 min) and failed to find a phase associated with q_T. However, a component due to state transitions would be unlikely under the conditions used by Schafer and Björkman (1989), where the material was kept for 1.5 hr in the light prior to fluorescence and photosynthesis measurements. They also suggest that both dark-relaxation phases are related to q_E, with the slower component tentatively related to zeaxanthin formation (Schafer and Björkman, 1989), constituting a possible protective mechanism against photoinhibitory damage (Demmig et al., 1987) and apparently synonymous with q_I.

Fluorescence Techniques

In conventional fluorometers the sample is irradiated with blue-green light that does not contain any of the wavelengths of the fluorescence emission. The fluorescence signal can be detected by a photodiode or a photomultiplier tube protected by suitable filters. For slow fluorescence transients, the signal can be recorded on a laboratory chart recorder, whereas resolution of the fast kinetics necessitates the use of an oscilloscope or transient recorder. Even though the instrumentation required is fairly simple, it must be remembered that this system measures fluorescence emission, which will be proportional to incident irradiation. A number of additional environmental factors will also alter the fluorescence transients (Renger and Schreiber, 1986), and there should be appropriate control over gas composition and temperature. For details of instrumentation, see Schreiber (1983). Suitable light sources and filters are described in Chapter 4.

Modulated Fluorescence

In conventional instruments, difficulties arise because the light source inducing fluorescence cannot contain any of the wavelengths of the emitted light. With this type of fluorometer, it is not possible to use white light to drive photosynthesis or the longer wavelengths of red and far-red light, which could otherwise be used, for instance, to examine state transitions. This problem can be overcome using a weak modulated light in conjunction with a system that selectively detects the fluorescence emitted at the same frequency and phase as the modulated source. With this technique, light of any wavelength can be used to induce photosynthesis, whereas the detection system only processes the signal that is due to the weak modulated source. The pulsed source is of a very low irradiance $(1-2 \ \mu mol \cdot m^{-2} \cdot s^{-1})$ and is generally incapable of generating variable fluorescence on its own. Available instruments can accurately detect the modulated signal despite large variations in background irradiance (Schreiber et al. 1986). Because the modulated source is of a constant irradiance, the modulated fluorescence signal induced by the actinic source will be constant once the background irradiance is high enough to saturate (close) all of the reaction centers. F_m will therefore be independent of the irradiance of the continuous saturating light source, unlike conventional fluorometers. Early instruments (Duysens and Sweers, 1963) were not commercially available and were too bulky and expensive to be used in general physiological or ecological studies. New technical developments have, however, resulted in the production of compact, relatively inexpensive instruments with a considerably wide range of uses (Bolhar-Nordankempf et al., 1989). Not only have these instruments been used to probe early photochemical events (Schreiber, 1986), but they are being used increasingly in ecophysiological studies to examine the relationship between slow fluorescence transients and photosynthesis. In conjunction with a high-output pulsed source, they can be used to generate a Kautsky curve and to estimate photochemical and nonphotochemical quenching components (see below). In all of these measurements, care must be taken to ensure that the modulated source only generates F_0 fluorescence. In material grown in low light, for instance, even an irradiance as low as 2 $\mu mol \cdot m^{-2} \cdot s^{-1}$ may generate some variable fluorescence. An estimate of the "true" F_0 can be obtained by varying the modulated light output to obtain the maximum F_v / F_m ratio and then using this irradiance for subsequent measurements (Horton and Hague, 1988). Alternatively, F_0 may be determined by using red actinic light to oxidize all of the intersystem carries, including Q_a, to ensure that all of the reaction centers are in an open state (Weis and Berry, 1987). Measurements made under transient or fluctuating conditions may necessitate frequent rechecking of the appropriate irradiance for obtaining the F_0 level.

The Pump-Probe Technique

Originally developed by Mauzerall (1972, 1980) to examine excited state lifetimes and fast fluorescence kinetics, the pump-probe technique has been used recently to examine the fluorescence emission characteristics of microalgae in the laboratory and in the field (Falkowski et al., 1986, 1988; Kolber et al., 1990). Other results using this method are reported in Mauzerall and Greenbaum (1989) and Geacintov and Breton (1987), although this method has not been used as extensively as other fluorescence techniques. However, it does have the advantage, which it shares with modulated techniques, that it can be used in continuous light. Cells are exposed to a weak (≈ 600 μmol$\cdot$m$^{-2}\cdot$s^{-1}) *probe* flash of several microseconds duration (from a laser or xenon arc lamp). This is subsequently followed by a high irradiance ($\approx 10,000$ μmol$\cdot$m$^{-2}\cdot$s^{-1}) *pump* flash and the fluorescence emission measured after a second probe flash (F_s). The pump flash irradiance can be varied to partially close or saturate the reaction centers to provide information on the response of fluorescence to different irradiances. The probe flash used alone (F_p) is thought to give a measure of fluorescence emission from antenna chlorophylls during steady-state CO_2 assimilation (apparently equivalent to F_0, but see below) and the high irradiance flash should induce maximum fluorescence (equivalent to F_m). The change in fluorescence yield (ΔF) is calculated from

$$\Delta F = (F_s - F_p)/F_p \tag{3.16}$$

It is not clear why ΔF is used rather than $F_s - F_p$, which would be equivalent to the variable fluorescence (F_v), or why this is normalized to F_p rather than F_s. Both F_p and F_s change significantly in response to irradiance (Falkowski et al., 1986), complicating the interpretation of fluorescence kinetics based on ΔF. When F_p and F_s are measured without background irradiance they are analogous to F_0 and F_m (Falkowski et al., 1986).

Using this technique, turnover times for PS2 can be obtained by varying the delay period between pump and probe flashes. Light-response curves for fluorescence can also be determined on the same sample, by varying the intensity of the pump flash. The normal procedure (Mauzerall, 1980; Falkowski et al., 1988; Mauzerall and Greenbaum, 1989) is to fit the curve to a cumulative one hit Poisson distribution. Interestingly, the variable fluorescence detected by the probe flash is generally an exponential function of irradiance, indicating a single population of PS2 centers and little interunit transfer of excitation energy, in contrast to the majority of results using conventional instrumentation (see above and Mauzerall and Greenbaum, 1989). The change in fluorescence yield normalized to the maximum change in fluorescence yield can then be related to an exponent called the apparent optical cross section for PS2 (σ_f, in the case of a measurement determined by fluorescence) and the pump flash irradiance.

$$\Delta F/F_m = 1 - \exp(\sigma_f E) \tag{3.17}$$

Although the pump-probe technique provides information on the redox state of Q_a during photosynthesis, which is not possible with conventional instrumentation, this technique has not been used to estimate photochemical and nonphotochemical quenching. The interpretation of fluorescence emission using this procedure is also very complex, and differences in the relationship between photosynthesis and fluorescence (Fig. 3.3), obtained with this methodology (Falkowski et al., 1986), compared to results using modulated techniques (Weis and Berry, 1987; Weis and Lechtenberg, 1989; Holmes et al., 1989), require further investigation.

77°K Fluorescence

At the temperature of liquid nitrogen (77°K) only photochemical reactions can take place, and the fluorescence emission characteristics are dependent almost entirely on the redox state of Q_a. No appreciable reoxidation of Q_a occurs during induction and temperature-dependent nonphotochemical processes are eliminated so that much simpler induction kinetics are observed. At 77°K, PS2 has two emission bands at 685 and 695 nm and PS1 has a band at 730 nm. Changes in the ratio of emissions at 685 or 695 nm, relative to those at 730 nm have been used to examine variations in the distribution of excitation energy between PS1 and PS2 (Butler, 1978). However, care must be taken in their interpretation as the 685 or 695 nm emission will be related to chlorophyll concentration, whereas the 730 nm emission will not be as dependent on pigment content because chlorophyll only absorbs weakly at longer wavelengths. Further details are given in Hipkins and Baker (1986).

Measurements of chlorophyll fluorescence at 77°K have been particularly useful as an indicator of environmentally induced damage to the photosynthetic apparatus despite the fact that the reoxidation of Q_a is blocked. With healthy nonstressed plant material, there are only small differences in F_v/F_m measured at 77°K, and variations in the F_v/F_m ratio due to environmental perturbations are often correlated with measurements of the photon yield of O_2 evolution measured by gas exchange techniques (see Björkman, 1987; Björkman and Demmig, 1987). These results are consistent with theoretical considerations (see below) and also indicate that environmentally induced changes in the oxidation state of Q_a are related to variations in light-limited photosynthesis. However, there are situations where reductions in the photon yield do not parallel variations in F_v/F_m, indicating that impaired electron transport either on the oxidizing or reducing side of the PS2 reaction-center complex may occur under some environmental conditions (Adams et al., 1990). In this case, where PS2 photochemistry does not limit CO_2 assimilation, measurements of F_v/F_m will not give a quantitative assessment of the state of the photosynthetic apparatus.

Relationship between Photosynthesis and Fluorescence

It has long been realized that there should be a complementary relationship between photosynthesis and fluorescence. McAlister and Myers (1941, cited in Butler, 1977), for instance, reported an inverse relationship between fluorescence emission and photosynthesis during induction and Delasome and coworkers (1959) described the relationship in *Chlorella* by the following formula.

$$P = c - aF \qquad (3.18)$$

where a and c are constants. Walker et al. (1983) have, more recently, demonstrated antiparallel relationships between photosynthesis and fluorescence emission during induction. However, until the development of modulated fluorometers, it was not possible to directly examine the relationship between photosynthesis and fluorescence under conditions of steady-state CO_2 assimilation.

The rate of photosynthetic electron transport (J) equals the product of the photon yield of electron transport (ϕ_s) and the rate of light absorption (A):

$$J = \phi_s A \qquad (3.19)$$

The photon yield of electron transport during steady-state assimilation is the product of the potential photon yield with all traps open (ϕ_p) and a photochemical quenching coefficient (q_Q) (see Weis and Lechtenberg, 1989).

$$\phi_s = \phi_p q_Q \tag{3.20}$$

If ϕ_p was constant, ϕ_s would be determined simply by the magnitude of q_Q. However, the evidence indicates that ϕ_p is variable and highly corre- lated with changes in nonphotochemical quenching (Weis and Berry, 1987; Holmes et al., 1989). This indicates that the steady-state photon yield is dependent on both q_Q and q_{NP}. Weis and Berry (1987) derived the following empirical relationship relating q_Q and q_{NP} to J.

$$J = q_Q A(b - mq_{NP}) \tag{3.21}$$

where b and m are experimentally-derived constants. Note the similarity of Eqs. 3.18 and 3.21. Strictly, fluorescence quenching is used to deter- mine the rate of photosynthetic electron transport. In turn, estimated rates of photosynthetic electron transport can be used to calculate rates of gas exchange by assuming a ratio of electrons transported (δe^-) to gross O_2 evolution (δO_2) or CO_2 consumption (δCO_2). Based on the Z- scheme for photosynthesis, the ratio of $\delta e^-/\delta O_2$ has a value of 4. The ratio $\delta e^-/CO_2$ is somewhat greater because various electron acceptors such as nitrate and sulfate compete with CO_2 for reducing power generated by photosynthetic electron transport, and because plant material is more reduced than carbohydrate. The relation between electron transport and CO_2 exchange can be calculated from the equations of Caemmerer and Farquhar (1981) which take into account the effects of respiration and photorespiration on net CO_2 exchange. Alternatively, electron transport can be estimated directly from the rate of gross O_2 evolution using mass spectrometry (Holmes et al., 1989).

Good agreement has been obtained between values of J determined from CO_2 exchange measurements and fluorescence kinetics for steady- state photosynthesis (Weis and Berry, 1987). However, uncoupling of CO_2 exchange from electron transport has been demonstrated under transient conditions in experiments conducted by Holmes et al. (1989) in which fluorescence quenching, gross O_2 exchange, and gross CO_2 exchange were measured for the chlorophyte Scenedesmus minutum (Fig. 3.4). Under conditions in which NO_3^- competed with CO_2 as an electron acceptor, CO_2 exchange greatly underestimated electron transport. How- ever, both fluorescence quenching and gross oxygen exchange gave comparable estimates of electron transport rate under both steady state and transient conditions (Fig. 3.3). Carbon assimilation was temporarily suppressed when NO_3^- was added to nitrogen-limited cells but returned to normal following uptake of all of the added NO_3^-. In contrast, both oxygen exchange and the electron transport rate were temporarily suppressed only after the added nitrate became depleted (Fig. 3.4).

One potential problem with this technique is the accuracy with which F_0 can be determined during the measurement period. Quenching of F_0, for instance (see Bilger and Schreiber, 1986; Horton and Hague, 1988) may result in erroneous values for F_v, q_Q, and q_{NP}, but does not appear to be a problem in the experiments of Holmes et al. (1989). A further uncertainty is the appropriate protocol to use for determinations of F_0. It is customary to determine F_0 in dark-adapted material (Holmes et al., 1989), although this would appear to be inappropriate for measurements conducted with continuous background irradiation, even under transient conditions. In contrast, Weis and Berry (1987) determined F_0 with background far-red light in order to fully oxidize the intersystem carriers and maintain all of the reaction centers in the open state.

We can also relate the photon yield of PS2 photochemistry to the fluorescence emission parameters F_v and F_m obtained during induction or under steady-state conditions. At the F_0 level, when all centers are open ($P = 1$),

$$\phi_{F0} = F_0/A = k_f/(k_f + k_d + k_t + k_p \cdot P) \qquad (3.22)$$

At the F_m level, when all of the reaction centers are closed ($P = 0$),

$$\phi_{Fm} = F_m/A = k_f/(k_f + k_d + k_t) \qquad (3.23)$$

as $\phi_{F_v} = \phi_{F_m} - \phi_{F_0}$, it can be shown that

$$\phi_{F_v}/\phi_{F_m} = F_v/F_m = k_p(k_f + k_d + k_t + k_p \cdot P) = \phi_p \qquad (3.24)$$

where ϕ_p is the photon yield of PS2 photochemistry. Accordingly, the photon yield of PS2 photochemistry measured indirectly by O_2-exchange techniques on whole plants should be related to the ratio F_v/F_m (but see Adams et al., 1990). Although measurements of 77°K fluorescence (Schafer and Björkman, 1989) largely confirm these predictions with terrestrial vascular plants, there are few if any comparable measurements with algae. Although correlations between CO_2-exchange measurements and F_v/F_m might be expected to be poor because of the possible influence of a number of downstream reactions on photosynthesis, there is often good agreement under many experimental conditions (Schafer and Björkman, 1989; Adams et al., 1990).

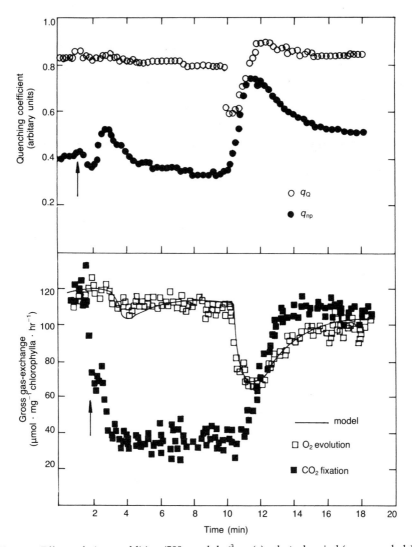

Fig. 3.4. Effects of nitrate addition (500 μmol.dm^{-3} on (a), photochemical (open symbols) and nonphotochemical (closed symbols) quenching coefficients, and (b) simultaneous changes in gross O_2 evolution (open symbols) and gross CO_2 consumption (closed symbols) with *Scenedesmus minutum*, measured by mass spectrometry. Also shown (solid line) is the rate of O_2 evolution estimated from the fluorescence quenching coefficients q_Q and q_{NP} calculated from Eq. 3.14. From Holmes et al. (1989).

In a recent analysis, Genty et al. (1989) showed that there is a direct linear relationship between the photon yield for CO_2 fixation (ϕ_{CO2}) and the fluorescence parameters F_m and F_s

$$\phi_{CO2} = [(F_m - F_s/F_m) - b]/m \qquad (3.25)$$

where b is the intercept and m is the slope. Under conditions in which F_0 (see above) and q_N are likely to exhibit large variations, this may provide a better estimate of electron transport than Eq. 3.21 (Keiller and Walker, 1990).

Chapter 4

Light Sources and Related Accessories

Often, there is little consideration of the type of light source required for gas exchange measurements, and the determining factor is generally availability rather than suitability. Obviously, the particular light source will depend on whether the lamps are to be used for growing algae or for direct measurements of photosynthesis or fluorescence. The former requires a number of low-cost lamps, with a low heat output, which can provide a relatively uniform Irradiance over a large area, whereas in most gas-exchange studies narrow beam or point sources are required. Failure to consider the merits of different light sources prior to experimental investigations can result in considerable methodologic problems due, for instance, to low outputs, excessive overheating, or poor spectral resolution. There are several factors to consider before choosing a particular light source. The incident irradiance will depend on the lamp power output, the distance between the lamp and the illuminated sample, and the type of optical system used. Filters or monochromators will drastically reduce the original light output. 12-V, 100-W tungsten-halogen lamps are adequate for most photosynthetic and absorption measurements (maximum irradiance $\approx$ 2,000 μmol·m^{-2}·s^{-1}, λ = 400–700 nm) using simple optical arrangements. More complicated systems using monochromators may require considerably higher output lamps. The source should have a continuous emission spectrum and be efficient in the photosynthetically active wavebands (λ = 350–720 nm). It should also have a long lifetime with a constant output that exhibits little variation with lamp age. Ideally, the lamp should have a minimal heat output or the housing should have a built-in cooling system.

Details of the theory, construction, and performance of a number of light sources can be found in the following references: Withrow and

Withrow (1956), Bickford and Dunn (1972), Jenkins and White (1976), Rurainski and Mader (1980), Smith and Holmes, 1984) and the technical handbooks of the major lamp manufacturing companies (e.g., Thorn Technical Handbook, 1989; Oriel Scientific, 1990; Newport, 1990; Philips Comprehensive Handbook, 1983*). Basically, the types of lamps available for photosynthetic and related measurements fall under four categories, incandescent sources, gas-discharge lamps, light-emitting diodes, and lasers.

Incandescent Lamps

Incandescent lamps are the most widely used light sources in photosynthetic studies. They consist of a coiled filament enclosed in a glass envelope and emit radiation due to thermal vibration of molecules when the filament is heated. The emission spectrum is continuous and the spectral distribution depends on the absolute temperature and emissivity of the surface. Partially evacuated lamps have a limited output (< 40 W) and the most useful sources contain argon and nitrogen gases under pressure, which prolong the lifetime of the lamp by reducing the evaporation of tungsten. These tungsten (quartz) halogen lamps can have outputs ranging from 10–1,000 W. Envelopes made of quartz can be operated at much higher temperatures than those made of hard glass, allowing the construction of smaller bulbs. In turn, the more compact construction allows the use of higher gas pressure, so that the lamp can be operated at higher temperatures, thereby increasing light output. For these reasons, it is a misconception to attribute high outputs to large lamp size.

Lamp life is governed by the length of time required before the continual evaporation of tungsten results in fracture of the filament. High-output lamps, such as those used in a projector, are more efficient than low-output lamps as the thicker filament allows higher operating temperatures to be used with greater output in the visible region of the spectrum. Even higher outputs can be achieved using dichroic reflectors, which preferentially reflect visible light and allow heat energy to escape through the reflector. Overheating should be prevented using forced ventilation or cooling fins, although air-cooling may not be sufficient for prolonged use. The bulb should be kept free from contamination by grease or particulate material as this can result in envelope failure. Except for linear tungsten-halogen lamps, the orientation of the source is generally unimportant, making them extremely versatile, although tungsten can

---

*Current catalogues from Philips do not include technical information. This is available on request. 1991 catalogue available.

be deposited on the envelope if the lamp fails to reach the optimum temperature when switched on for periods less than 30 s, making them unsuitable for measuring fast transients or the 'S' states of O_2 evolution.

One of the major disadvantages of tungsten-halogen light sources in photosynthetic studies is their low output in the blue region of the spectrum. This severely limits their usefulness in measurements of action spectra or spectral absorption. In addition, measurements made with these lamps may not be directly applicable to natural conditions of differing spectral quality. The use of dichroic reflectors marginally improves the spectral quality of the source as can operation of the lamp at a higher than recommended output, but this will reduce lamp longevity. New tungsten-halogen lamps have a quartz envelope with a molybdenum foil hermetic seal, which allows the filament to be operated at a higher temperature, thereby increasing the output in the blue spectral region.

Discharge Lamps

When high irradiances are required, incandescent lamps may provide an inadequate light output. Discharge lamps have no filament and light is produced by excitation of a gas with an electric current passed between two electrodes. Energy absorbed by the gas is reemitted in the form of light due to the resonant vibration of atomic structures. The gases used are normally mercury and sodium vapors. In the xenon and mercury-discharge lamps, the gas is contained within an inner quartz envelope that encloses the two electrodes, whereas short-wave emitters or fluorescent tubes only have one outer envelope. The wavelengths of light produced depend largely on the gas used, although the spectral quality can be modified by coating the glass envelope with a fluorescent power (phosphor).

Mercury-Discharge Lamps

These may contain mercury vapor together with halogen gas or metallic sodium. Lamps containing halogen gases are often called metal-halide lamps, but this leads to some confusion with tungsten-halogen sources, which are not discharge lamps. When operated at low gas pressures, the spectrum consists of several lines superimposed on the existing (continuous) mercury spectrum. Most of the light is emitted in the yellow and ultraviolet regions of the spectrum. Although the spectral quality can be improved by the addition of traces of halogen gases or metallic sodium, these low-pressure lamps are generally unsuitable for photosynthetic studies. Considerable improvements in spectral quality can, however, be achieved in mercury-discharge lamps containing gases at high pressures. This results in a higher output in the visible region and a broadening of the individual lines, so that the spectrum is almost continuous. Short or compact-arc versions of these lamps, where arc refers to the distance between the electrodes (1–14 mm), are the most suitable for photosynthetic studies and make excellent point sources. Although they require efficient forced ventilation, the bulb itself should not receive cool air directly and should be wired to switch off at the same time as the lamp (contrast with the recommendations for the xenon-arc lamp below).

Xenon-Discharge Lamps

These lamps work on the same principle as mercury-discharge lamps, but the mercury gas is replaced with xenon gas or a mixture of mercury and xenon. Short-arc and compact-arc versions are also available. In contrast to mercury lamps, they provide an almost constant light output between 400 and 800 nm. They are more stable than mercury-arc lamps and have a greater longevity, but have a higher infrared component. These lamps make excellent pulsed sources for gas-exchange and fluorescence studies and, unlike mercury discharge lamps, can be switched on when hot. Most lamps have to be operated within 20% of the vertical, although various ancillary optical arrangements are now available, which give some flexibility in beam direction. Correct focusing is essential to optimize light output and to avoid permanent damage to the lamp in both mercury and xenon-arc lamps. Forced cooling is essential and should continue for at least 5 min after the lamp is switched off (contrast with the mercury lamp). Xenon lamps also generate ozone because short-wave length radiation below 200 nm produces O_3, which can be a potential safety problem. Low power ozone-producing lamps should be operated in well-ventilated conditions, whereas O_3 produced by lamps of higher output should be vented to the exterior. New lamps have modified glass envelopes that are opaque to radiation below 240 nm and are largely ozone-free.

Fluorescent Lamps

Although these light sources are often categorized separately, they are really only low-pressure mercury-discharge lamps in which the arc is generated within a cylindrical glass tube internally coated with a fluorescent powder. Electrodes are located at each end of the tube, which can be of variable length and which typically contains argon or krypton with traces of mercury. An electric current passed through the tube initially excites the mercury vapor. Deexcitation of the mercury vapor then results in the production of ultraviolet light, which is in turn converted to visible light by the phosphor coating on the tube wall. The spectral quality of the emitted light is determined by the type of phosphor used and a wide variety of these compounds are now available. Modern developments in fluorescent tube design and performance have resulted in significant increases in output by using either reflectors or by twisting the glass envelope into a spiral shape (Vita-Lite, Power Twist, Duro–Test Corporation, U.S.A.) and with phosphors that result in a continuous spectrum similar to natural daylight. Typically, however, fluorescent tubes have a low light output and emit light that is predominantly in the orange/red regions of the spectrum. Their cylindrical construction makes them unsuitable as point sources for use in gas exchange studies. However, they have a low heat output and are extremely efficient and stable, with a long lifetime, making them ideally suited for controlled growth studies on algae, either on their own or with supplementary light.

Light-Emitting Diodes

Light-emitting diodes (LEDs) are based on conventional diodes and consist of two fused semiconductors, the p and n type junctions, operated with a forward bias current. In conventional diodes, movement of electrons from the n to the p type junction results in the loss of thermal energy. However, the introduction of large quantities of impurities, called *dopants*, to the largely silicon-based semiconductor, can result in the production of visible light. The spectral quality of the emitted light depends on the type and amount of dopant used and the operating temperature. The output is a function of the applied current and the operating temperature. Generally the maximum current that can be used continuously is approximately 100 mA, although larger currents (5 A) can be used during pulsed operation. Power consumption is low (approximately 40 mW) so that they can be operated from very small batteries. Schreiber (1983) has suggested a number of potential advantages of using LEDs in fluorescence studies. (1) LEDs quickly reach maximum output when the current is switched on and can be rapidly switched on and off, making them ideal for pulsed operation with no requirement for a shutter system. (2) Their light output can be modulated without the requirements for a mechanical chopper making them suitable for lock-in amplification systems. (3) Due to their small size, LEDs can be packaged into self-contained units for irradiating large areas. In addition, LEDs have a low heat output, are easy to operate, inexpensive, and have a long-lifetime. In the past, a major drawback to their use in photosynthetic studies was their low radiation output caused by repeated self-absorption of the generated light. Consequently, only a small fraction of the generated light is emitted from the diode. This can be overcome, to some extent, by coating the outer covering with a material of low refractive index and by packaging a number of LEDs into a diode array. During pulsed operation considerably higher outputs can be obtained, and it is possible to saturate PS2 reaction centers with a single millisecond flash (Schreiber, 1983), because higher currents can be used. Individual LEDs can produce continuous irradiances > 50 μmol·m^{-2}·s^{-1} over small areas (Schreiber, 1983), whereas commercially available diode arrays can produce irradiances in the red region of the spectrum (900 μmol·m^{-2}·s^{-1}), which are saturating for many C_3 plants (see Walker, 1989). In new LEDs, variations in irradiance can be achieved by varying the current without any change in spectral output, thereby eliminating the need for neutral density filters. Improvements in semiconductor technology are also likely to result in further increases in light output, and LEDs should have increasing use in phycological research. Instruments are already available for instantaneous spectral absorption measurements using an array of filtered diodes, each with a different peak spectral output, and this equipment is almost certain to be used more extensively in studies on algae where variations in spectral light quality are required.

Lasers

With ordinary light sources, each atom emits photons independently and the intensity of the source is the sum of the contributions of each individual atom. As the light emitted by individual atoms is out of phase (incoherent), both destructive and constructive interference can occur and the *actual* output is less than the potential output of the source. In the laser (an acronym for light amplification by stimulated emission of radiation), all of the light emitted from different atoms is in phase (coherent), and the total intensity is proportional to the square of the number of atoms. Consequently, there is an enormous increase in output from lasers in comparison to conventional lamps, and irradiances several times that of full sunlight can be achieved. The light waves are kept in phase by a phenomena called stimulated emission. As an excited electron returns to a lower energy level a photon is released that stimulates another excited electron to return to its ground state in unison with the original photon, rather than decaying spontaneously. The original photon stimulates the emission of an additional coherent photon.

All lasers operate on the same fundamental principles. The key to their operation is the maintenance of the majority of atoms in an excited and relatively long-lived (metastable) state, through a continuous input of energy. The energy input can be supplied via an electric current or from a high-output light source. Typically, lasers are constructed of a long cylindrical tube, which encloses the lasing material, with reflective end walls one of which is semitransparent. Reflections from the end walls increase the light pathlength within the tube and excites more atoms to emit photons in phase. Due to the tubular construction only photons that travel along the axis of the tube will be emitted from the semitransparent window, and the beam is composed of waves traveling in exactly the same direction. Lasers differ only in the type of material used in the cylinder cavity, but this results in quantitative and qualitative variations in light output. Contrary to some suggestions, the basic laser does not emit a single spectral band and additional bands of reduced intensity are present. Conversely, lasers do not provide broad-band (λ = 400–700 nm) radiation and cannot be used to investigate plant responses to natural light fields.

The Ruby Laser

This laser consists of a long cylindrical crystal of synthetic ruby containing Al_2O_3 impregnated with Cr^{3+}. Flashes from a xenon lamp are used to produce excited Cr^{3+} atoms (called optical pumping) and photons are reflected at each end of the tube by mirrorlike silvered end walls, one of which is partially transparent. The spectral output is confined to several lines in the red region of the visible spectrum and, together with its pulsed operation, limits its usefulness in photosynthetic studies. In addition, the beam diverges and does not provide a uniform light source.

The Helium-Neon Laser

A continuous output can be obtained from a helium-neon laser. This consists of a glass tube, with silvered mirrors at each end, containing a mixture of helium and neon at low pressure. A high-voltage power supply is used to excite the helium gas, which in turn transfers energy to the neon atoms. The Ne atoms will lase at 633 nm and also at several wavelengths in the infra-red region of the spectrum. This laser has a less divergent beam than the ruby laser.

The Carbon Dioxide-Nitrogen Laser

This operates in a similar way to the helium-neon laser, with the N_2 atoms exciting the CO_2, which lases. Lasers of this kind can provide 10 kW of continuous output and have the added advantage of being operable in a pulsed mode, with an even greater output. Unfortunately, current lasers emit only infrared light and no visible light.

The Semiconductor Laser

This uses a *p–n* junction, usually made of gallium arsenide with a dopant of zinc and operates on a similar principle to that already described for the LED sources. In contrast to LEDs, it is important in this system to retain light within the diode to obtain the greatest signal amplification. This is achieved using a sandwich of semiconductor materials of graded refractive indices. These lasers are normally compact, require low operating voltages, and are easily modulated, but they require high current densities and the resulting high heat loads require an efficient cooling system for continuous operation. Solid-state diode lasers will operate with outputs up to 50 W, with a broad range of wavelengths, making them especially versatile for photosynthetic studies. New semiconductor lasers that are used in supermarket check-out systems are inexpensive and may provide the type of low-cost system that is required in many phycologic studies.

Tunable/Dye Lasers

The light energy emerging from a laser is a mixture of several wavelengths coinciding with variations in the energy gaps of different excited states. To obtain monochromatic light from a laser source various devices must be attached to or used in conjunction with the basic light source. Conventional optical accessories such as narrow-band filters or monochromators can be used (see the next section). In addition, there are a number of modifications that can be made to the laser to enable the separation of discrete spectral lines. The usual procedure is to place a prism with one silvered face near the transparent end of the tube. The prism will refract light of different wavelengths at different angles. By changing the orientation of the prism, a selected wavelength will strike the silvered face at a right angle and will be reflected back into the laser tube. Lasing will therefore only occur at this particular wavelength. By adjusting the prism, different wavelengths can be obtained, although the range of available wavelengths will be quite small.

Dye lasers use molecules of organic dyes, which emit a large number of closely spaced spectral bands within a defined spectral region. Different dyes are available to cover an extremely wide range of wavelengths. In combination with a prism, spectral lines can be isolated within a spectral band covered by a particular dye. These lasers seem to have considerable potential in many areas of photosynthesis research, but have been little used, as yet. There is little information, for instance, on spectral variations in photoinhibition of photosynthesis, because of an inability to achieve the high irradiances that are required. This should now be possible using dye lasers.

Optical Filters

Optical filters are required in virtually all gas-exchange and fluorescence measurements to modify or restrict the quantity or quality of electromagnetic radiation. The choice of a particular filter depends on a number of factors, including the size of the area to be irradiated, the waveband required, its transmission characteristics, and resistance to thermal shock. All filters operate on the basis of either selective absorption or transmission, or by constructive and destructive interference effects. Filters that transmit on one side of a certain wavelength and little on the other side, are called cutoff filters, whereas those that transmit in one or more distinct wavebands are called bandpass filters. Bandpass filters are characterized by their maximum percentage transmission (peak transmission), by the wavelength at which maximum transmission occurs (peak wavelength or wavelength maximum), and the bandwidth (or half-bandwidth), which is the width of the transmitted waveband usually at 50% of the peak transmission. Because of asymmetry in the shape of the transmission curve, further information may also be given for the bandwidths at 90%, 10%, 1%, 0.1%, and 0.001% of the transmission peak in order to fully describe the filter characteristics. Some filters have secondary transmission peaks in addition to the main peak, and these may need to be eliminated using the appropriate filter combination. A number of other terms are often used to characterize particular filters. The cut-on or cut-off point is the point where the transmission starts to increase from a low value up to a maximum and is usually taken as 5% of the maximum transmission. However, a cut-off filter is normally used in a restricted sense to refer to filters that absorb shorter wavelengths. In some applications, sharp cut-off or cut-on points may be required to prevent any overlap between the selected wavebands. *Blocking filters* largely eliminate certain wavebands by either absorption or reflection, and the blocking range is the spectral range over which these wavelengths are reduced. The degree of blocking refers to the extent, normally a percentage, to which the waveband is reduced or blocked. The rejection ratio or degree of rejection is the ratio of the peak transmission to the transmission within the blocked waveband, expressed as a percentage.

Absorption Filters

Absorption filters are generally broad-band filters with bandwidths varying from 30–250 nm. Perhaps the simplest absorption filters are those made from a suspension of salts or dyes suspended in a liquid (often water). A solution of $CuSO_4$ in water, for instance, is routinely used to block infrared. Others are made from colored glass or from dyes suspended in gelatin, acetate, or plastic. Gelatin filters of the "Wratten" (Kodak) type have relatively sharp absorption bands, and have been used extensively in fluorescence measurements to protect the photodiode from measuring wavelengths of light outside the fluorescent band. However, they are fragile and easily damaged by moderate temperatures (> 40°C) and high humidities. Acetate (Cinemoid) filters are stronger than gelatin, although they have larger bandwidths. Plastic filters are rigid enough to be self-supporting and suitable for producing light fields over large areas or as narrow waveband sources in photosynthetic studies, as long as the temperature is maintained below 80°C. Glass absorption filters are particularly useful where there are high head loads (250–500°C). They are often used as cut-off filters, which transmit over large regions of the spectrum, but decrease to zero transmittance over a small wavelength range. Long-wave pass filters transmit longer wavelengths while rejecting short wavelengths, whereas short-wave pass filters transmit short wavelengths and reject longer wavelengths. Long- and short-wave pass filters can, however, only be used in the transmission mode (contrast with comparable interference filters). Variations in filter orientation alter their percentage transmission, but otherwise their optical characteristics remain unchanged and, with the exception of cut-off filters, these characteristics are largely independent of alterations in temperature. Cut-off filters, in contrast, do exhibit some shift toward longer wavelengths at high temperatures and all absorption filters deteriorate with age. Although bandpass filters can be constructed from combinations of glass filters, they have a low transmittance and are generally inferior to interference filters.

Interference Filters

These filters transmit relatively narrow-band radiation and are constructed from layers of semitransparent mirrors separated by cavities or spaces. Multiple reflection between the layers results in constructive interference of some wavelengths, which are transmitted and destructive interference of other wavelengths, which are reflected. Variations in the number and spacing of the layers and their reflectivities determines the peak wavelength transmitted and the bandwidth. The peak wavelength varies linearly with changes in temperature, with typical values for the −50°C to +70°C range of 0.015 nm/°C at $\lambda = 400$ nm. Temperature fluctuations of > 5°C should also be avoided because of the possibility of permanent damage caused by thermal shock. As the filter characteristics vary regularly with wavelength, these variations can be used to tune the filter to a specific wavelength within the permissible temperature range, using temperature controllers. Accurate temperature control will also reduce the possibility of damage due to fluctuations in ambient temperature. Filter orientation can also seriously affect the optical characteristics of interference filters, with a shift toward shorter wavelengths as the angle of incidence increases. This characteristic can also be used to tune an interference filter to a specified wavelength, but may result in considerable reductions in transmittance and an increase in the bandwidth at large angles of incidence. Long- and short-wave pass interference filters are also available and, as they use interference phenomena rather than absorption, they can be used in either the transmittance or reflectance mode. Other interference filters can be used, as alternatives to absorption filters, to transmit certain regions of the visible spectrum while reflecting all the others. As the filters are effectively absorption free, high light transmission is ensured. Unlike colored glass or gelatine filters, interference filters do not deteriorate with time, but are susceptible to damage by moisture if the sealing cement is damaged.

Heat-Reflecting and Heat-Transmitting Filters or Mirrors

These filters are used extensively in conjunction with broad-band light sources to isolate infrared radiation from the visible wavelengths and to maintain surface temperatures within reasonable tolerance limits. Heat-reflecting and heat-transmitting filters consist of a single dielectric coating on a polished Pyrex glass surface. Hot mirrors reflect, rather than absorb, heat, giving them some advantages over heat-absorbing glass filters. Cold mirrors reflect visible light but transmit the infrared and are used primarily at an angle of 45°C. Ultraviolet cold mirrors are also available. Infrared-suppression filters transmit in the visible region of the spectrum and, because of a combination of a reflective coating and an absorbing glass, they can completely block the infrared.

Neutral Density Filters

These consist of a metallic alloy coating deposited on high-quality glass or polished fused silica and are used to attenuate light with minimal changes in spectral quality. They are categorized by their different optical densities (i.e., absorbance) and can be used in series to obtain the required output. However, due to multiple reflections between filters the actual output may be less than the calculated value based on the optical densities of the filters. This can be overcome by tilting the filters a few degrees away from the normal, without changing their rated optical properties. Most neutral density filters do not have the same transmittance at all wavelengths, although many have a fairly constant transmittance to broad-band ($\lambda = 400–700$ nm) radiation, and these types should be used for photosynthetic studies. The filters should always be used with the reflective face toward the source to minimize the thermal load on the filter components and only used in conjunction with heat filters or mirrors. Aluminum gauze or netting (1.4 × 1.4-mm grid, 0.28-mm wire diameter mosquito netting) has been reported to have a negligible effect on the spectral quality of the transmitted radiation (see Smith and Holmes, 1984) and would seem to be an ideal heat-resistant and inexpensive filter for covering large areas, or for use in laboratory light sources.

Monochromators

For many applications in photosynthesis research, a system for providing a variable range of monochromatic wavelengths is required (e.g., for determination of action spectra). The simplest way to generate a range of known wavelengths of light is to use a suitable light source and a series of narrow-band interference filters. Although many custom-built systems based on interference filters exist (see Smith and Holmes, 1984, Lewis et al., 1986), a wide range of high-performance monochromators are now available commercially. Before choosing a particular monochromator, there should be some consideration of the most suitable light source to use. With any light source, suitable lenses are needed for focusing the beam, and filters are required to remove excessive amounts of infrared radiation, all of which reduces the transmitted flux in addition to any losses within the monochromator itself. As an example, losses due to lenses can be of the order of 8% and heat-absorbing filters can reduce the transmitted flux from 20 to 30%. There will be additional losses in the forward direction, even if auxiliary mirrors are used, because of less than 100% reflection at the mirror surfaces. Even if the lenses are covered with (expensive) antireflective coatings, reducing losses to 2% rather than 8%, total losses will still be > 60% of the original flux. With modern monochromators, maximum transmittances approach 45%, reducing the flux out of the exit slit of the monochromator to only 27% of its original value. This is also a maximum value; in the blue spectral regions, the transmitted flux from the monochromator may be only 20%, resulting in a reduction in the output flux to only 12% of the value originating directly from the source. Even lower figures will be found with grating monochromators because of greater losses within the optical components of the system. Consequently, relatively high-output lamps are required for any measurements requiring the use of monochromators. In absorption measurements, sensitive detectors, such as photomultipliers, can compensate for the small signal. However, during measurements of spectral variations in photosynthesis, a high output is essential to induce measurable responses from an O_2 electrode or an infrared gas analyzer. The use of high-output lamps (1,000 W) that are deficient in the shorter wavelengths, such as a tungsten-halogen source, may still result in poor spectral resolution in the $\lambda = 350$–450 nm waveband. For the equivalent power (1,000 W), a xenon discharge lamp has a two- to eightfold higher output in the visible region, with even greater differences in the photosynthetically active 350–400-nm waveband. Past complications with the use of xenon lamps (see above) and their prohibitive cost, which is well in excess of tungsten-halogen sources, has limited their use in photosynthesis research, but they are the preferred light source for all spectral measurements.

All monochromators reduce the transmitted flux considerably; nevertheless, it is still important to optimize the system to maximize the flux out of the monochromator, even if a high-output source is used. Firstly, the appropriate source should be chosen, with a high output in the spectral regions of interest. Secondly, the maximum possible slit width should be used, and the size of the source should be matched as close as possible to the slit dimensions. Thirdly, as much light as possible should enter the entrance slit within the acceptance angle of the monochromator. For high-output systems, a small high-output source should be imaged on the entrance slit using lenses or mirrors. In grating monochromators, the appropriate grating should be used for the particular wavelength region required. Although monochromators are supposed to provide light in a narrow restricted waveband, stray white light may escape from the exit slit because of scattering of the unwanted portions of the spectrum within the housing and losses due to reflections from the optical components. These effects are particularly significant when the actual signal is low and is an additional problem when tungsten-halogen lamps are used. Stray light can be reduced by using blocking filters, by narrowing the input spectrum with a broad- or narrow-band filter, or by using two monochromators in series, but this will result in considerable reductions in the transmitted flux.

Basically, there are two types of monochromator: the filter monochromator and the grating monochromator. The filter monochromator consists of a continuously variable interference filter. All are restricted to measurements within the visible region of the spectrum and cannot be used for measurements over the whole photosynthetically active ($\lambda = $ 350–720 nm) waveband. Modern instruments are compact, have a high transmission, a wide collecting cone, and an efficient rejection of stray light using a built-in blocking filter. Supplementory infrared blocking filters may, however, be required. Wavelength resolution is low, and the bandpass is typically no better than 20 nm, so that detailed spectral measurements are not possible.

The grating monochromator, in contrast, separates spectral regions by the diffraction of a beam of polychromatic light with a diffraction grating. A typical diffraction grating consists of a flat surface on which are ruled a large number of parallel grooves, separated by distances that approach the wavelength of light, and covered with a reflective material. By changing the orientation of the grating relative to the direction of the incoming flux, diffracted photons of different wavelengths can be made to pass out through the exit slit. Clearly, the scattering of unwanted wavelengths within the system can be a problem and needs to be minimized (see above). Losses within the system will depend on the efficiency with which the grating diffracts light, and this depends on the wavelength region being examined. Alterations in the angles of the grooves and the shape or depth of the grating rulings are used to optimize their efficiency, and interchangeable gratings are available for different spectral regions. Spectral resolution is high, with a bandpass of < 0.5 nm, but the output is low and stray light losses are greater than with filter monochromators.

Chapter 5

Measuring Photosynthetic Pigments

The concentration of photosynthetic pigments is commonly used as a measure of the amount of plant material in a sample (Cullen, 1982), and photosynthesis rates are frequently normalized to chlorophyll *a* concentration (Jassby and Platt, 1976). In addition, assessments of pigment composition provides qualitative information on the physiological condition of natural samples by providing information on the presence of breakdown products (Hallegraeff and Jeffrey, 1985) and provides information on the taxonomic composition of algal populations *via* an analysis of accessory pigments (Jeffrey, 1976; Liaaen-Jensen, 1977; Jeffrey and Hallegraeff, 1987a, b; Wright and Jeffrey, 1987; Hooks et al., 1988). Estimations of pigment content and complement are also the basis of quantitative assessments of the relationship among light harvesting, photosynthesis, and growth (Mann and Myers, 1968; Bidigare et al., 1989).

Chlorophyll *a* is the only photosynthetic pigment found in all oxygen-evolving photoautotrophs and so provides a convenient and unambiguous measure of algal abundance (Cullen, 1982). However, the ratio of chlorophyll *a* to algal biomass ranges from 0.1% to 5% of dry weight, and varies in response to changes in irradiance, nutrient availability, and temperature (Kirk, 1983; Geider, 1987). In addition to chlorophyll *a*, algae contain accessory pigments that often account for most of the *in vivo* absorption (Sathyendranath et al., 1987). These pigments include chlorophylls *b* and *c*, carotenoids, and phycobiliproteins.

Extraction of Chlorophylls and Carotenoids

Chlorophylls *a* and *b* consist of a magnesium porphyrin complex with a long hydrophobic side chain (Jeffrey, 1980), whereas chlorophyll *c* is actually a chlorophyllide that lacks the phytol tail. Carotenoids are long polyisoprenoids having conjugated double bonds with unsaturated un-substituted cyclohexane rings at each end of the molecule (Goodwin, 1980). There are two classes of carotenoids: carotenes, which are hydro-carbons, and xanthophylls, which contain oxygen atoms in their terminal rings. Chlorophylls can be characterized spectrophotometrically as hav-ing relatively sharp absorption maxima in both the blue and red spectral regions. The carotenoids, in contrast, usually have a broad yellow absorp-tion band, perhaps with several local maxima within this broad band.

Chlorophylls and carotenoids can be extracted in organic solvents: acetone has been the solvent of choice for oceanographers, although investigators who study macroalgae and freshwater phytoplankton pre-fer the more polar solvent methanol, which provides a higher extraction efficiency. Macroalgal samples are blotted dry, cut into small pieces, and extracted by grinding tissues in a mixture of solvent and an abrasive such as sand, whereas microalgae are commonly concentrated by filtration, followed by homogenization in a solvent using a tissue grinder (Jensen, 1978). Solubilized pigments are separated from residual material by cen-trifugation or filtration. The homogenization step is sometimes omitted in analyses of microalgae in order to decrease the processing time. Spe-cifically, a sample can be concentrated onto a filter, placed in a vial containing solvent, and allowed to extract overnight in the cold at 4°C or in a deep freeze at −20°C. Phinney and Yentsch (1985) recommend direct injections of samples into 100% acetone (to obtain a final concentration of 85%) for pigment extraction of microalgae.

Incomplete pigment extraction may be encountered with less labor-intensive procedures, particularly when chlorophytes and cyanobacteria are present in samples extracted with 90% acetone (Shoaf and Lium, 1976; Stauffer et al., 1979). Complete extraction from microalgal samples can be obtained with either hot 100% methanol or a 1:1 mixture of dimethyl sulfoxide (DMSO) and acetone (Shoaf and Lium, 1976; Stauffer et al., 1979). Complete extraction of pigments from macroalgae presents more problems, and an initial extraction with DMSO following by extrac-tion with acetone or methanol has become a method of choice for ex-tracting pigments from recalcitrant species (Duncan and Harrison, 1982). Plant physiologists have found that N_0N-dimethylformamide (DMF) is an effective solvent for extracting chlorophylls a and b from vascular plants (Moran and Porath 1980). This solvent may also prove useful for macroalgal samples.

Chlorophylls and carotenoids are labile compounds that may decom-pose when exposed to light, acids, or oxygen. Precautions must be taken to avoid chemical degradation and photooxidation: these include working at low temperature under subdued illumination or darkness and using deoxygenated solvents (Jensen, 1978). Addition of H_2S to methanol is commonly recommended, and $MgCO_3$ is often added to the sample prior to extraction to neutralize organic acids, which may be released when tissues are disrupted.

The principal breakdown products of chlorophylls *a* and *b* are pheophytins, chlorophyllides, and pheophorbides. Acidification of chlorophyll leads to pheophytin production through removal of magnesium from the porphyrin ring. Chlorophyllide is produced by the removal of the phytol chain from chlorophyll *a* by chlorophyllase. Pheophorbide is produced by the removal of both magnesium and phytol. Pheophytins and pheophorbides may be present at large concentrations in natural samples as a result of zooplankton grazing, senescence, and/or decomposition. Chlorophyllide may be produced during sample extraction because chlorophyllase remains active in aqueous organic solvents. In addition, pheophytin or pheophorbide may be produced if organic acids are not neutralized during extraction. Chlorophyllase is active even in 90% acetone, and extraction in 100% acetone has been recommended to inhibit chlorophyllase activity (Jeffrey, 1974; Jeffrey and Hallegraeff, 1987a). Even this may not be sufficient, and Suzuki and Fujita (1986) suggest a brief exposure to heat (65°C) to completely inactivate chlorophyllase.

Quantification of Chlorophylls and Carotenoids

Pigments are commonly identified by their absorption spectra, and routine methods for determining pigment concentrations in solvent extracts rely on spectrophotometric measurements. Several different expressions are used to quantitatively describe the absorption characteristics of pigments. Much of the older data is reported as specific extinction coefficients designated E_{1cm}, where 1 cm refers to the pathlength of the spectrophotometer cell used to measure absorbance. More recently, molar and weight-specific absorption coefficients have become the preferred units. The following relationship can be used to convert among these different expressions:

$$A^* = A_m/M = E_{1cm}10 \tag{5.1}$$

where: A^* is the $\log_{10}$ weight-specific absorption coefficient ($m^2 \cdot g^{-1}$), A_m is the molar absorption coefficient ($m^2 \cdot mol^{-1}$), E_{1cm} is the specific extinction coefficient ($dm^3 \cdot g^{-1} \cdot cm^{-1} = 0.1\ m^2 \cdot g^{-1}$), and M is the molecular weight ($g \cdot mol^{-1}$) of the pigment. The weight of pigment in a pure sample (such as would be obtained following separation by thin-layer chromatography) can be calculated from the specific absorption coefficient as follows:

$$W/V = D/(A^*l) \tag{5.2}$$

where W is the weight of pigment (in grams), V is the volume of solution (in cubic meters), D is the absorbance measured in a spectrophotometer cell with pathlength l (in meters), and A^* is the specific-absorption coefficient (Table 5.1).

By substituting the specific-absorption coefficient for the red absorption maximum of chlorophyll a into Eq. 5.2, one can calculate the concentration of chlorophyll a in extracts from rhodophytes and cyanobacteria, which lack accessory chlorophylls b and c. However, Eq. 5.2 cannot be used where absorption bands of several pigments overlap. Series of equations have been developed to allow estimation of chlorophylls a and b or chlorophylls a and c in unialgal samples, as well as for the simultaneous estimation of all three chlorophylls in mixed assemblages (Table 5.2). The trichromatic equations of Jeffrey and Humphrey (1975) are based on reliable estimates of specific absorption coefficients and yield satisfactory results when applied to mixed algal assemblages (Lorenzen and Jeffrey, 1980).

Determination of chlorophyll concentrations by spectrophotometry in samples that contain degradation products is problematic. Chlorophyllides have nearly identical absorption spectra to undegraded chlorophylls (Table 5.1) and thus can only be separated by chromatography. Pheophytins and pheophorbides show reduced absorption when compared to chlorophylls (Table 5.1): this is associated with the loss of magnesium from the center of porphyrin ring of chlorophyll. A crude estimation of chlorophyll a and pheophorbide a concentrations is possible by measuring the absorption at 665 nm before and after acidification of a 90% acetone extract (Lorenzen, 1966). The method does not allow discrimination of chlorophyll a from chlorophyllide a, nor pheophytin a from pheophorbide a. In addition, there is no spectrophotometric technique for simultaneously estimating chlorophylls a, b, and c and their degradation products.

Samples for spectrophotometric pigment determination can be readily obtained from macroalgal thalli, which have chlorophyll a concentrations of > 100 mg·m^{-2}. However, relatively large volumes of natural phytoplankton assemblages, in which chlorophyll a concentrations can be < 0.1 µg·dm^{-3}, may need to be concentrated to obtain enough material for spectrophotometric analysis. A major improvement in routine estimation of chlorophyll concentration in open ocean waters was provided by the introduction of a fluorometric technique (Yentsch and Menzel, 1963;

Table 5.1. Specific absorption coefficients, molecular weights and absorption maxima of chlorophylls, chlorophyll *a* degradation products and selected carotenoids.

Pigment	Solvent	λmax (nm)	$A*(\lambda$max) $(m^2 \cdot g^{-1})$	Molecular Weight $(g \cdot mol^{-1})$
Chlorophyll *a*	Diethyl ether	423	12.8[a]	893
	Diethyl ether	662	9.85[a]	
	Acetone	666	8.90[b]	
	90% Acetone	664	8.77[b]	
	80% Acetone	664	7.68[f]	
	Methanol	664	7.14[f]	
Chlorophyllide *a*	90% Acetone	664	12.7[d]	615
Pheophytin *a*	90% Acetone	667	5.12[d]	869
Pheophorbide *a*	90% Acetone	667	7.42[d]	591
Chlorophyll *b*	Diethyl ether	453	17.5[a]	907
	Diethyl ether	643	6.1[a]	
	90% Acetone	647	5.14[c]	
	80% Acetone	647	4.70[f]	
	Methanol	652	3.86[f]	
Chlorophyll c_1	Acetone/pyridine	447	34.9[e]	611
	Acetone/pyridine	630	3.93[e]	
	Acetone	629	3.92[c]	
	90% Acetone	630	4.26[c]	
Chlorophyll c_2	Acetone/pyridine	447	32.2[e]	609
	Acetone/pyridine	630	3.73[3]	
	Acetone	631	3.72[c]	
	90% Acetone	631	4.26[c]	
β-carotene	Petroleum ether	453	25.92[b]	537
γ-carotene	Petroleum ether	462	31.00[b]	537
Diadinoxanthin	Acetone	453	23.0[b]	659
Fucoxanthin	Acetone	449	16.0[b]	659
Lutein	Acetone	445	25.0[b]	569
Peridinin	Acetone	466	13.4[b]	630
Violaxanthin	Acetone	442	24.0[b]	601
Zeaxanthin	Acetone	452	23.4[b]	569

References are as follows: (a) Mauzerall (1977); (b) Jensen (1978); (c) Jeffrey and Humphrey (1975); (d) Lorensen and Jeffrey (1980); (e) Jeffrey (1972); (f) Porra et al. (1989). For a more complete list of absorption maxima and E_{1cm} for carotenoids see Davies (1976).

Holm-Hansen et al., 1965) that allows up to a 100-fold increased sensitivity over spectrophotometric methods. A comparison of spectrophotometric and fluorometric determinations of chlorophyll *a* from ocean waters indicated agreement between the two techniques to within ± 20% (Holm-Hansen et al., 1965). Measurement of fluorescence before and after acidification of a sample enables the determination of chlorophyll *a* and

Table 5.2.   Spectrophotometric equations for estimating chlorophyll concentrations.

Jeffrey and Humphrey's (1975) equations for 90% acetone extracts.

Chlorophylls a and b
$$Chla = 11.93\ D_{664} - 1.93\ D_{647}$$
$$Chlb = -5.50 D_{664} + 20.36\ D_{647}$$

Chlorophylls a and c
$$Chla = 11.47\ D_{664} - 0.4\ D_{630}$$
$$Chlc = -3.73 D_{664} + 24.36\ D_{630}$$

Chlorophylls a, b and c
$$Chla = 11.85\ D_{664} - 1.54\ D_{647} - 0.08\ D_{630}$$
$$Chlb = -5.43 D_{664} + 21.03\ D_{647} - 2.66 D_{630}$$
$$Chlc = -1.67 D_{664} - 7.60\ D_{647} + 24.52 D_{630}$$

Lorenzen's (1967) equations for 90% acetone extracts.

Chlorophyll a and pheophytin a
$$Chla = 26.7\ [D_{665}(o) - D_{665}(a)]$$
$$Pheoa = 26.7\ [1.7 D_{665}(a) - D_{665}(o)]$$
where $D_{665}(o)$ is the absorbance of a 90% acetone extract and $D_{665}(a)$ is the absorbance of the 90% acetone extract following acidification to pH 2.7.

Jeffrey and Haxo's (1968) equations for 100% methanol extracts.

Chlorophylls a and c
$$Chla = 13.8\ D_{668} - 1.3\ D_{635}$$
$$Chlc = -14.1\ D_{668} - 67.3\ D_{635}$$

Porra et al. (1989) equations for 100% methanol extracts.

Chlorophylls a and b
$$Chla = 16.29\ D_{665} - 8.54\ D_{652}$$
$$Chlb = -13.58\ D_{665} + 30.66\ D_{652}$$

All concentrations are given in units of $\mu g \cdot cm^{-3}$ where D is the absorbance in a 1-cm pathlength cell.

pheophorbide a, but, unfortunately, as routinely employed, the fluorescence technique does not allow the determination of chlorophylls b and c.

Unlike the spectrophotometric methods, which give an absolute measure of concentration based on the specific absorption coefficients of the pigments under investigation, the fluorescence methods give a relative measure of pigment concentration and must be calibrated with solutions of known composition. Fluorescence techniques were originally calibrated against acetone extracts of log-phase algal cultures in which chlorophyll a concentration was determined by a spectrophotometric procedure. However, the relationship between chlorophyll a and fluorescence depends on the accessory pigments that are present, and thus the calibration can vary with the algae used to obtain the pigment extract. To overcome this problem, many investigators prefer calibration against purified chlorophyll *a*, which provides an objective standard, even though this may lead to a bias in the estimation of chlorophyll *a* concentration.

Holm-Hansen et al. (1965) recognized that errors in estimating chlorophyll *a* by the fluorometric technique could result from interference by other pigments that also fluoresce red when excited by blue light. Using data of Lorenzen and Jeffrey (1980), Trees et al. (1985) were able to provide an equation for predicting the effects of other pigments on the concentration of chlorophyll *a* determined by fluorometry. The equation relating the chlorophyll *a* concentration obtained by the fluorometric technique of Holm-Hansen et al. (1965) to the pigments present is:

$$C_m = 0.974 \text{ (chlorophyll } a) - 0.282 \text{ (chlorophyll } b)$$
$$+ 0.380 \text{ (chlorophyll } c) + 0.0793 \text{ (pheophorbide } a) \qquad (5.3)$$
$$+ 1.392 \text{ (chlorophyllide } a)$$

where C_m is the measured chlorophyll *a* concentration and the parentheses are used to indicate the true concentrations of the indicated pigments (Trees et al., 1985). Thus, the presence of significant chlorophyll *b* in a sample will lead to underestimates of the chlorophyll *a* concentration, whereas the presence of significant amounts of chlorophyll *c* or the degradation products chlorophyllide *a* or pheophorbide *a* will lead to overestimates of the chlorophyll *a* concentration.

Chromatographic Separation of Pigments

To completely characterize the pigment composition of a sample, it is necessary to chromatographically separate pigments prior to identification and quantification. Quantitative thin-layer chromatography was first used on algal pigments by Jeffrey (1968) and was subsequently improved (Jeffrey, 1981; Vesk and Jeffrey, 1987). More recently, reverse-phase high-performance liquid chromatography (HPLC) techniques have been developed for routinely separating chlorophylls, chlorophyll degradation products, and carotenoids (Schwartz and von Elbe, 1982; Mantoura and Lewellyn, 1983). Mantoura and Lewellyn's (1983) method required a 20-min run to separate pigments using a two-mobile-phase solvent system on a 250 × 5 mm C-18 column (5-μm particle size). Increasing column diameter and solvent flow rate Trees et al. (1986) achieved separation of chlorophylls and their degradation products in a 10-min run. Recently, de los Rivas et al. (1989) describe a new three-phase solvent system and a 10 × 8 mm radial compression column, which allows separation of terrestrial vascular plant pigments within a 10-min run. Widespread application of HPLC and thin-layer chromatography in recent years has led to the identification of previously unknown pigments and to the recognition of the importance of these pigments in various algal groups. A consideration of this research is beyond the scope of this review, and further discussion will be limited to a comparison of HPLC and other methods for measuring chlorophyll a.

In general, HPLC and other techniques lead to comparable estimates of chlorophyll concentration, although individual estimates may be greatly at variance. For example, Murray et al. (1986) compared two HPLC techniques, a fluorometric method (Strickland and Parsons, 1972) and a spectrophotometric method (Jeffrey and Humphrey, 1975) for estimating chlorophyll a in 29 plankton samples with chlorophyll a concentrations ranging from 0.02 to 30 $\mu g \cdot dm^{-3}$. The chlorophyll a peak accounted for only 57% of the spectrophotometer response and 72% of the fluorescence response. However, including chlorophyll allomers in the comparison led to a one-to-one correspondence of spectrophotometric and HPLC methods. In sediments with chlorophyll a and pheopigment concentrations of 1–30 $\mu g \cdot g^{-1}$ dry weight, Riaux-Gobin et al. (1987) found that the fluorescence measurements underestimated chlorophyll a and pheopigments by 30% to 40% relative to the HPLC measurements.

Trees et al. (1985) found that fluorometric determinations of chlorophyll a were in error by from -68% to $+53\%$ when compared with values obtained from a HPLC method. They found that errors in the fluorometric determination varied between stations within a given oceanographic region, and between regions. When data from various oceanographic regions were combined, chlorophyll a determined from HPLC and fluorometric techniques were highly correlated with a correlation coefficient of 0.961 (Trees et al., 1985). Despite the high correlation obtained when measurements over a large range of chlorophyll a concentrations (< 10–40,000 $ng \cdot dm^{-3}$) were compared, individual estimates of chlorophyll a by the fluorometric technique were as much as three times lower or 1.5 times greater than the true value.

The increased discrimination available from HPLC analysis of phytoplankton pigments has also proven useful in ecological investigations. Aside from providing qualitative information on the major taxa present, a quantitative estimate of the relative contribution of phytoplankton with different accessory pigments to total chlorophyll a can be obtained from the application of a multiple regression analysis to quantitative pigment determinations (Gieskes et al., 1988). Changes in the concentrations of accessory pigments can also be used to characterize the growth and loss rates of different phytoplankton taxa in grazing experiments (Burkhill et al., 1987).

Extraction and Quantification of Phycobiliproteins

Phycobiliproteins are water-soluble light-harvesting pigment proteins found in the rhodophytes, cryptomonads, and cyanobacteria. The major classes of phycobiliproteins are phycocyanin (PC), phycoerythrin (PE), allophycocyanin (APC), and phycoerythrocyanin (PEC) (Gantt, 1981). Within each of these classes are phycobiliproteins with several related spectral forms. The phycobiliproteins consist of open-chain tetrapyrrole chromophores (referred to as bilins or phycobilins) covalently bound to proteins. The main chromophore types are (1) phycocyanobilin, found in PC and APC, (2) phycoerythrobilin found in PE, and (3) phycourobilin found in some PEs (Gantt, 1981; Cohen-Bazire and Bryant, 1981).

There is no routine method generally used for determining phycobili-protein concentrations. However, as in the case of lipophilic pigments, most methods employ extraction of pigment molecules, followed by spectrophotometric or fluorometric determination of chromophore concentration (Moreth and Yentsch, 1970; Stewart and Farmer, 1984). Extraction of phycobiliproteins can be achieved by mechanical and/or enzymatic cell disruption, which releases phycobiliproteins into aqueous solution. Laboratory cultures of microalgae and cyanobacteria are commonly disrupted by repeated cycles of freeze-thaw treatment and passage through a French press (Alberte et al., 1984), or by prolonged sonication (Sigelman & Kycia, 1978). Phytoplankton can be concentrated onto membrane filters and disrupted by mechanical shearing and lysozyme treatment (Stewart and Farmer, 1984), although recoveries may not be complete in recalcitrant organisms. Particularly troublesome in this regard are the small coccoid cyanobacteria, which are resistant to mechanical disruption. Wyman (unpublished) sought to overcome this problem when measuring phycoerythrin in marine *Synechococcus* spp. by measuring in vivo fluorescence of samples resuspended in 50% glycerol. Glycerol effectively uncouples phycoerythrin from photosynthetic energy transduction enhancing fluorescence by up to 15-fold relative to untreated cells. A constant maximum fluorescence yield under these circumstances allows calibration against extracted phycoerythrin. The technique is simple, rapid and has been used for routine measurements at sea (Wyman, unpublished). Variations in the magnitude of the package effect (Chapter 5) may preclude a simple quantitative relationship between glycerol enhanced fluorescence and phycoerythrin in large cyanobacteria, but this

Table 5.3.   Spectral characterization of various phycobiliproteins.

Phycobiliprotein	Absorption λ_{max}	Fluorescence Emission λ_{max}	Molecular Weight $(g \cdot mol^{-1})$
C-phycocyanin	620	650	120,000
R-phycocyanin	553, 615	640	130,000
Cryptomonad phycocyanins			
phycocyanin-615	585, 615	—	—
-630	585, 630	—	—
-645	585, 620, 645	—	—
phycoerythrocyanin	568, 590	610	103,000
allophycocyanin B	618, 673	680	98,000
allophycocyanin I	654	680	145,000
allophycocyanin II	650	680	105,000
allophycocyanin III	650	680	105,000
R-phycoerythrin	498, 542, 565	578	260,000
B-phycoerythrin	498, 545, 563	575	260,000
b-phycoerythrin	545, 563	575	—
C-phycoerythrin	565	578	—
C-phycoerythrin I	555	578	—
C-phycoerythrin II	540, 568	578	—
Cryptomonad phycoerythrins			
Phycoerythrin-544	544, 565	—	—
-555	555	—	—
-568	508	—	—
C-phycoerythrins			
Synechococcus WH7803	500, 545	560	—
Synechococcus WH8018	551	570	—
Trichodesmium	498, 557	—	—
Gloeobacter violacea	501, 564	574, 577	—

Compiled from summaries in Gantt (1981), Kirk (1983), and Alberte et al. (1984).

effect should be insignificant for the *Synechococcus* spp. examined by Wyman (unpublished).

Although different phycobiliproteins can be identified by characteristic absorbance and/or fluorescence emission spectra (Table 5.3), the diversity of these pigments complicates quantification. An indication of the range of specific absorption coefficients that have been obtained is given in Table 5.4. Bennett and Bogorad (1973) and Kursar and Alberte (1983) developed trichromatic equations that account for spectral overlap for estimating phycocyanin, allophycocyanin, and phycoerythrin concentrations from absorption measurements in aqueous extracts (Table 5.5). These equations are specific to the organisms under investigation, and,

Table 5.4. Specific absorption coefficients (a^*, with units of $m^2 \cdot g^{-1}$ pigment) of some phycobiliproteins in phosphate buffer.

Biliprotein	λ_{max}	Molecular Weight	a^*	Reference
C-phycocyanin	618	—	0.724	Kursar and Alberte (1983)
C-phycocyanin	615	224,000	0.770	Siegelman and Kycia (1978)
C-phycocyanin	620	—	0.650	Stewart and Farmer (1984)
R-phycocyanin	615	273,000	0.660	Siegelman and Kycia (1978)
allophycocyanin	650	—	0.603	Kursar and Alberte (1983)
allophycocyanin	650	90,000	0.650	Siegelman and Kycia (1978)
B-phycoerythrin	565	290,000	0.820	Siegelman and Kycia (1978)
B-phycoerythrin	545	—	0.878	Stewart and Farmer (1984)
C-phycoerythrin	565	226,000	1.250	Siegelman and Kycia (1978)
C-phycoerythrin	550	—	1.250	Stewart and Farmer (1984)
R-phycoerythrin	565	298,000	0.810	Siegelman and Kycia (1978)
R-phycoerythrin	565	—	0.802	Myers et al. (1980)
phycoerythrin	545	—	1.260	Stewart and Farmer (1984)
phycoerythrin	548	—	1.050	Alberte et al. (1984)
phycoerythrin	551	—	0.800	Alberte et al. (1984)
phycoerythrin (M)	492	264,000	1.05	Ong and Glazer (1988)
	543	—	0.432	

The designations B, C, and R indicate spectroscopically identifiable pigments. The designations were originally based on the higher taxon from which a pigment was isolated (specifically, B designated a pigment extracted from the Bangiales, C designated a pigment identified in the Cyanobacteria, and R designated a pigment identified in the Rhodophyta). This taxonomic distinction no longer holds, but the prefixes B, C, and R have been retained (Cohen-Bazire and Bryant, 1982).

for maximum accuracy, individual equations need to be derived for each particular application (Siegelman and Kycia, 1978).

In Vivo Fluorescence for Estimating Chlorophyll *a*

Measuring the red fluorescence of chlorophyll *a* when algae are illuminated with blue light has become a standard method of determining microalgal abundance in natural samples (Lorenzen, 1966) and unialgal cultures (Brand and Guillard, 1981). Although subject to some uncertainty because of variations in fluorescence emission (Falkowski and Kiefer, 1985), in situ determination of relative phytoplankton abundance is possible, and absolute abundance can be obtained if these measurements are calibrated against extracted chlorophyll *a*.

At physiological temperatures, in vivo chlorophyll *a* fluorescence arises predominantly from photosystem 2 and is characterized by a peak in the

Table 5.5.   Equations used to estimate phycobiliprotein content in aqueous extracts.

Bennet and Bogorad (1973) equations for *Fremyella diplosiphon*
$$PC = 0.187\ D_{615} - 0.0888\ D_{652}$$
$$APC = 0.196\ D_{652} - 0.0409\ D_{615}$$
$$PC = 0.104\ D_{562} - 0.0505\ D_{615} - 0.0395\ D_{652}$$

Kursar and Alberte (1983) equations for *Neoagardhiella bailyei*
$$PC - 0.166\ D_{618} - 0.108\ D_{650}$$
$$APC = 0.200\ D_{650} - 0.0523\ D_{618}$$
$$PE = 0.169\ D_{498} - 0.0086\ D_{615} - 0.0018\ D_{650}$$

All concentrations are given in units of $mg \cdot cm^{-3}$ where D is the absorbance in a 1 cm pathlength cell. The concentrations calculated with these equations include the amount of chromophore and associated protein.

emission spectrum at about 685 nm (Chapter 3). Fluorescence at shorter wavelengths in phycobiliphytes and cryptomonads arises largely from inefficiencies in energy transfer within phycobiliproteins. Although in vitro fluorescence of pigments in organic or aqueous solvents depends only on the structure of the pigment molecule, in vivo fluorescence also depends on localized interactions with specific proteins and lipids, together with cell structural and metabolic variations. As a consequence, the ratio of in vivo fluorescence to extracted chlorophyll *a* can vary by approximately an order of magnitude (Loftus and Seliger, 1975). The relationship between fluorescence emission and chlorophyll *a* concentration can be considered as the product of the *in vivo* chlorophyll *a*-specific absorption coefficient (a^*) (see Chapter 6), the photon yield of fluorescence (ϕ_f) and the incident irradiance (E),

$$F_{chl} = a^*\phi_f E/2.303 = a^*\phi_f E \tag{5.4}$$

where F_{chl} is the chlorophyll-*a*-specific fluorescence emission. The absorption coefficient, a^*, is influenced by factors that affect light absorption (Chapter 5), whereas ϕ_f may be affected by a range of physiological variables (Kiefer, 1973).

The chlorophyll *a*-specific fluorescence emission has been shown to have a size dependence in microalgal cultures (Loftus et al., 1972) and natural phytoplankton assemblages (Alpine and Cloern, 1985). Specifically, F_{chl} is found to vary approximately by a factor of five between picoplankton (2 μm diam) and net plankton (> 20 μm diam) (Loftus et al., 1972; Alpine and Cloern, 1985), consistent with expectations based on the relationship between a^* and cell size that are predicted by the package effect (Chapter 5). In addition to a dependence on cell size, F_{chl} varies by a factor of two to four due to physiological responses to irradiance and nutrient availability (Kiefer, 1973; Loftus and Seliger, 1975). Chloroplast aggregation in response to high irradiances resulted in decreases of both absorption (λ 440 nm) and fluorescence in the large diatom *Lauderia borealis* (Kiefer, 1973).

Changes in ϕ_f may also influence F_{chl}. For example, F_{chl} tends to be higher in stationary-phase cultures than in exponential cultures (Loftus and Seliger, 1972), and is related to a reduction in the photon yield of photosynthesis under nutrient-starved conditions (Kiefer, 1973; Cleveland and Perry, 1987). Up to an order of magnitude diel variation in F_{chl}, in which F_{chl} is highest at night and lowest during the day (Kiefer, 1973; Loftus and Seliger, 1975), may also reflect changes in ϕ_f. Significantly, the diel variation in F_{chl} is opposite to that observed for photosynthesis rates (Chapter 6), suggesting complimentary variations in the photon yield of fluorescence and photosynthesis (Chapter 3).

Fluorescence Microscopy

Epifluorescence microscopy has only recently become a standard tool for examining microalgal samples despite the introduction of fluorescence techniques to phycological research over 30 years ago. In early applications, Wood (1955) employed transmitted light fluorescence microscopy in plankton studies, and Jones (1974) used epifluorescence microscopy to identify living algae on the surfaces of stones. Epifluorescence microscopy has many advantages over transmitted-light, phase-contrast, or interference-contrast microscopy for examining plankton samples (Murphy and Haugen, 1985; Tsuji, Ohki, Fujita, 1986). Small organisms can be identified as algae by their red fluorescence, whereas cyanobacteria, cryptomonads, and rhodophytes can be characterized by distinctive orange or yellow fluorescence, enabling the separate enumeration of photosynthetic from heterotrophic forms. Use of this technique indicates that between 10 and 90% of oceanic nanoplankton (approximately 2–20 μm diam) are photosynthetic, with autotrophs typically accounting for 50% of total counts (Caron, 1983; Geider, 1988). In another application of fluorescence microscopy, Lessard and Swift (1986) found that approximately 40% of oceanic dinoflagellate taxa lack chlorophyll a, indicating a large phagotrophic component in a group traditionally regarded as autotrophic.

Quantitative determinations of the chlorophyll or phycobiliprotein content of single cells may be possible when a fluorescence microscope is used as a microspectrofluorometer. However, variations in the photon yield of fluorescence that hamper the in vivo determinations of chlorophyll concentration of cell suspensions will also influence measurements made with single cells.

Use of Derivative Spectra in Pigment Analysis

Faust and Norris (1982, 1985) describe an in vivo spectrophotometric technique for determining chlorophylls *a*, *b*, and *c* on intact phytoplankton. The technique rests on the observation that taking the second derivative of the absorption spectrum leads to considerably greater detail than is evident in the original data (Butler, 1979). The second derivative yields the rate of change of the slope of the absorption spectrum and so accentuates differences in curvature associated with in vivo absorption peaks. In its original application to microalgae (Faust and Norris, 1982), regression analysis was used to find the optimum wavelengths for determining chlorophylls *a*, *b*, and *c* from second derivative absorption spectra of samples concentrated on glass-fiber filters. The method was calibrated with a mixture of microalgae from major taxonomic groups: that is, the cyanobacterium *Anabaena flos-aquae*, the green *Chlorella* sp., the cryptomonad *Cryptomonas ovata*, and the dinoflagellate *Prorocentrum mariae-lebour-*

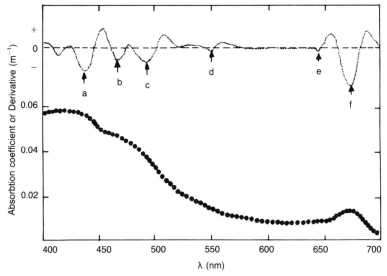

Fig. 5.1. Spectral variations in the absorption coefficient of a natural phytoplankton assemblage and their second derivative spectra. The letters a–f denote the derivative positions of the major pigments, chlorophyll *a* (a) , chlorophyll *c* (b), phycobiliproteins (c,d), chlorophyll *b* (e) and the red absorbing form of chlorophyll *a* (f), which are not clearly identifiable in the original in vivo absorption spectra. Redrawn from Bigidare et al. (1989).

iae. As employed by Faust and Norris (1982), the technique is strictly correlative, requiring calibration of the in vivo spectra against chlorophyll *a*, *b*, and *c* concentrations determined in solvent extracts of the same samples. Optimum wavelengths for determining these chlorophylls from second derivative spectra differed from the wavelengths of peak absorption in vivo, presumably because of overlap of absorption bands of chlorophylls and other pigments and absorption flattening, which causes nonlinearities at wavelengths of maximum absorption (Chapter 5).

Bidigare et al. (1989) used second and fourth derivative analysis to locate in vivo absorption peaks of pigments in open ocean phytoplankton (Fig. 5.1). Peaks for chlorophylls *a*, *b*, and *c*, as well as for phycoerythrobilin and phycourobilin were apparent in these spectra, despite the lack of features in the original in vivo absorption spectra, which were dominated by the contribution of detrital material. In contrast to the empirical approach of Faust and Norris (1982; 1985) described above, Bidigare et al. (1989) chose to examine the relations between the most prominent peaks in the derivative spectra (which corresponded to known in vivo absorption maxima) and the concentrations of these pigments.

Chapter 6

Light Utilization and Optical Properties of Algae

In many aquatic systems, the capture of light energy is regarded as the primary factor controlling plant productivity. Measurements of the optical characteristics of algae are essential for estimating the efficiency of photosynthetic light utilization, interpreting the in vivo significance of quantitative and qualitative changes in light harvesting pigments, and for determining the role of algae in modifying the transmission and reflection of light by natural waters. Despite this importance, optical measurements are made infrequently, largely because of difficulties associated with partitioning attenuation into components of absorption and scattering. In particular, measuring light absorption is complicated when scattering (reflection, refraction, and diffraction) is the dominant process that contributes to light attenuation. However, techniques are available for determining the optical properties of unicellular and multicellular algae, and it is likely that further progress in areas such as remote sensing and aquatic productivity will proceed from a more complete understanding of the absorption and scattering properties of algal cells and tissues.

The angular dependence of the scattered light for homogeneous spheres can be predicted from Mie theory (Mie, 1908) or the simpler anomalous diffraction theory (Hulst, 1957). Scattering by unicellular algae tends to conform to these predictions (Morel and Bricaud, 1981), although there is some evidence for a disproportionally high percentage of wide-angle scattering in some algal cells (Geider and Osborne, 1987). This wide-angle scattering could be due to internal inhomogeneities, such as organelles and other cell structures.

The interaction of a cell with a beam of light is commonly quantified in terms of the cross-sectional area of the incident beam that is modified by the presence of the cell. For scattered photons, this area is called the scattering cross section of the cell (sQ_b), where s is the cross-sectional area of the cell and Q_b is the efficiency factor for scattering. Dividing the scattering cross section by the geometric cross section gives the efficiency factor for scattering (Q_b). Similar expressions can be used for the absorption cross section (sQ_a) and the efficiency factor for absorption (Q_a). Since the total amount of light attenuated is the sum of scattered and absorbed light, an efficiency factor for attenuation (Q_c) can also be determined:

$$Q_c = Q_a + Q_b \qquad (6.1)$$

Interestingly, Q_c, Q_a, or Q_b can theoretically exceed unity, and a cell can influence the light field over an area that is greater than its geometrical cross-sectional area. This is most likely associated with diffraction and interference phenomena around the periphery of the cell (Hulst, 1957). In practice, however, only measured values for Q_c and Q_b exceed 1, whereas values for Q_a are significantly < 1 (Table 6.1) indicating that "edge" effects increase scattering more than absorption.

Considerably less information exists on the absorption and scattering characteristics of multicellular algae and theoretical descriptions of the optical properties of multicellular tissues have rarely been attempted. Phenomenologic theories have a limited application and do not provide quantitative descriptions of the influence of tissue dimensions or structures on absorption and scattering (see Osborne and Raven, 1986). Thick thallii (> 1 mm) of many macrophytes have spectral absorption and reflectance properties that are similar to the leaves of terrestrial vascular plants (see Ramus, 1978; Forbes and Canny, 1981; Gausman, 1985; Dring and Lüning, 1985; Osborne and Raven, 1986). Absorption, reflection and transmission show only small variations with wavelength across the visible spectrum, and differences in pigment content or complement have little effect on light absorption. The flatness of the absorption spectrum between 400 nm and 700 nm, which is largely independent of pigment content, is normally ascribed to increases in the pathlength of light within the tissue because of multiple internal reflections (Ramus, 1978; Ruhle and Wild, 1979; Osborne and Raven, 1986). On this basis, structural features appear to have a greater influence on light absorption than do variations in pigment content. There are, however, few detailed studies on the optical characteristics of macroalgae and many measurements ignore the reflectance component. The optical properties of macrophytes measured in a fixed position, perpendicular to the incident light, may also be modified *in situ* because of water movements or varia-

Table 6.1. Experimentally determined measurements of a number of optical parameters.

Parameter	(Minimum) (Maximum) $(m^2 \cdot mg^{-1})$		Parameter	(Minimum) (Maximum) (μm^2)		Parameter	(Minimum) (Maximum) (dimensionless)	
a^*	0.004	0.043	sQ_a	0.024	7.2	Q_a	0.020	0.65
b^*	0.044	0.60	sQ_b	39	61	Q_b	0.278	2.5
c^*	0.09	0.63	sQ_c	41	72	Q_c	0.308	3.05
b^*_b	0.075×10^{-3}	0.075×10^{-2}	sQ_{bb}	41×10^{-3}	151×10^{-2}	Q_{bb}	1.4×10^{-2}	6.5×10^{-2}

Values are given as the range of reported values. Data taken from Bricaud et al. (1983, 1988), Bricaud and Morel (1986), Haardt and Maske (1987), Table 4 of Maske and Haardt (1987), Morel (1987), Morel and Bricaud (1981), and Osborne and Geider (1989). Where reported, cell diameters and intracellular chlorophyll a concentrations varied from 1.5 to 26.5 µm and from 0.3 to 6.16 kg·m⁻³, respectively. All absorption parameters are for $\lambda = 675$ nm, whereas the scattering parameters are for values at $\lambda = 540$–590 nm. For a definition of symbols see Table 6.3.

tions in the underwater light field. Some algae found appressed to reflective substrates may also show an enhanced absorption in comparison to freely suspended material. With thin multicellular tissues (< 1 mm thick), absorption often shows a greater dependence on pigment content (Ramus, 1978), although not all macrophytes with tissues comprising only two to three cell layers show this dependence (Osborne, unpublished). Absorptance of thick tissues typically equals 90% of incident light and is independent of pigment content. It is difficult to understand the reasons for the increased synthesis of accessory pigments in multicellular algae in response to reductions in light level or changes in spectral quality, if the function of these pigments is solely a means of increasing the total amount of light absorbed.

There is considerable interest in internal light environments within the photosynthetic tissues of terrestrial plants and their functional significance (Fukshansky, 1981; Seyfried and Fukshansky, 1983; Terashima and Saeki, 1983). One of the important consequences of the highly scattering characteristics of plant tissues is that the magnitude of the light gradient is much greater than would be predicted from measurements made with a nonscattering system. The Kubelka-Munk theory predicts (Seyfried and Fukshansky, 1983) and experiments verify (Vogelman and Bjorn, 1984) that backscattering results in a higher irradiance just below the tissue surface closest to the light source than the incident irradiance. Differences in gradients of light within leaves have been correlated with qualitative differences in the response of photosynthesis to light (Terashima, 1986; Leverenz, 1987; 1988) and correlated with changes in tissue biochemistry (Terashima and Inoue, 1985). Whether gradients of similar magnitude exist in macrophytic algae awaits experimental investigation. Studies using small fiber-optic probes (Vogelman and Bjorn, 1984; Vogelman et al., 1988) could be used in measurements on multicellular algae, as could phenomenologic descriptions based on reflection and transmission measurements (Seyfried and Fukshansky, 1983). Recent measurements within cyanobacterial mats indicate that the light gradients may be of a similar magnitude to those reported for terrestrial plant species (Jorgensen and Des Marias, 1988).

Characterization of the Light Environment

The major terms used to characterize the underwater light environment are shown in Table 6.2. Since the electromagnetic spectrum is continuous, it is necessary to define the wavelength range used when measuring the light field. Often a broad so-called "photosynthetically active" wave-

Table 6.2. Radiation terms used to characterize the incident light field.

Term	Symbol	Definition	Units
Radiant flux	Φ	The flux of radiant energy per unit time	W ($J \cdot s^{-1}$), photon$\cdot s^{-1}$
Radiant intensity	I	Radiant flux per unit solid angle	$W \cdot steradian^{-1}$, photon$\cdot s^{-1} \cdot steradian^{-1}$
Field radiance	L_f	Radiant flux per unit area per unit solid angle	$W \cdot m^{-2} \cdot steradian^{-1}$, photon$\cdot m^{-2} \cdot s^{-1} \cdot steradian^{-1}$
Surface radiance	L_s	Radiant flux per unit projected area per unit solid angle	$W \cdot m^{-2} \cdot steradian^{-1}$, photon$\cdot m^{-2} \cdot s^{-1} \cdot steradian^{-1}$
Irradiance	E	Radiant flux per unit area of surface	$W \cdot m^{-2}$, photon$\cdot m^{-2} \cdot s^{-1}$
Downward irradiance	E_d	Irradiance on the upperface of a horizontal surface	$W \cdot m^{-2}$, photon$\cdot m^{-2} \cdot s^{-1}$
Upward irradiance	E_u	Irradiance on the lower face of a horizontal surface	$W \cdot m^{-2}$, photon$\cdot m^{-2} \cdot s^{-1}$
Net downward irradiance	**E**	Difference between the upward and downward irradiance	$W \cdot m^{-2}$, photon$\cdot m^{-2} \cdot s^{-1}$
Scalar irradiance	E_o	Integral of the radiation distribution over a full 360° (4π)	$W \cdot m^{-2}$, photon$\cdot m^{-2} \cdot s^{-1}$
Downward scalar irradiance	E_{od}	Integral of the radiation distribution over the upper 180° (2π) hemisphere.	$W \cdot m^{-2}$, photon$\cdot m^{-2} \cdot s^{-1}$
Upward scalar irradiance	E_{ou}	Integral of the radiation distribution over the lower 180° (2π) hemisphere.	$W \cdot m^{-2}$, photon$\cdot m^{-2} \cdot s^{-1}$
Irradiance reflectance or Irradiance ratio	R	Ratio of the upper to the downward irradiance	Dimensionless
Photon flux density	E^*	The number of photons incident on a unit area of surface.	photon$\cdot m^{-2} \cdot s^{-1}$

band (λ = 400–700 nm) is used. Although it is clear that algae can photosynthesize at wavelengths of < 400 nm and > 700 nm (Halldal, 1964), the 400–700 nm waveband is employed because of instrument limitations that preclude extending the upper and lower wavelengths (McCree, 1972). For many applications it is desirable to resolve spectral variations in the light field, in which case the wavelength under consideration must be designated explicitly. It should also be noted that photon or energy units are often used synonymously, although the conversion between the two is not constant and will depend on the spectral quality of the incident light. The terms *irradiance* and *photon flux density* indicate the incident flux per unit area on either emergent or submerged photosynthetic structures. They are also used to estimate the flux on an imaginary surface within the suspending medium. In contrast, *scalar irradiance* measures the integrated flux about a point (180° or 360°). Other terms in Table 6.2 describe the angular distribution of radiation and the extent to which radiation is attenuated with water depth. The angular distribution of radiation can be used to estimate the effectiveness of the incident flux for rigid planar photosynthetic surfaces. Measurements of downward (downwelling) irradiance give depth-dependent estimates of the availability of light for photosynthesis and are a consequence of the absorption and scattering properties of the photosynthetic structures and the surrounding medium. Upward (upwelling) radiation is generally a very small proportion of the downward flux and is due to backscattering by suspended cells, other particulate matter, reflectance from submerged multicellular tissues or the bottom, and water itself.

The basic sensor now used to measure light flux (λ = 400–700 nm) is a filtered silicone photodiode with a cut off at 400 and 700 nm and gives an integrated measure of the irradiance over this waveband. Details of the construction of these sensors are given in Woodward and Sheehy (1983) and in the manufacturers literature. Basically, there are three types of sensor/collector (Fig. 6.1), the flat-plate cosine-corrected collector (surface area = πr^2), the 2π collector (surface area = $2\pi r^2$), and the 4π collector (surface area = $4\pi r^2$) (Ramus, 1985). These differences in collector area are important because the amount of light collected in a *diffuse* light field will be directly proportional to the collector's surface area. Although both the cosine-corrected and the 2π collector measure light over a 180° field of view, they differ because the cosine-corrected collector has a reduced response to light at low solar elevations. Ideally, this response should obey the Lambert cosine law, where the light level is equal to the product of the flux on a surface normal to the incident beam and the cosine of the angle of incidence. Most collectors have responses that agree well with theory, except at very low angles of incidence. In contrast, the 2π and 4π collectors respond equally to all photons

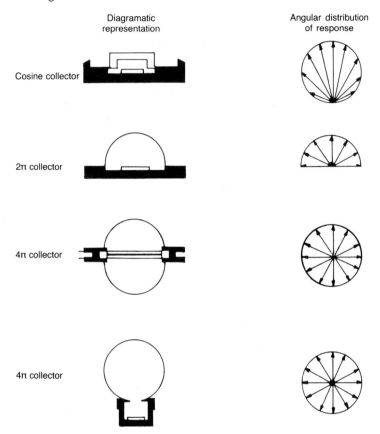

Fig. 6.1. Illustration of the types of light detectors used to measure incident irradiance together with the angular distribution of the detector response. Cosine collectors are used to measure horizontal irradiance and exhibit a reduced response to light originating from wide angles; 2π collectors are used to measure hemispheric irradiance over an $180°$ cone and 4π collectors are used to determine spheric or scalar irradiance over a full $360°$. Both 2π and 4π collectors should exhibit an equal response to light of the same flux (photons per second), irrespective of its angular distribution. For further details see text. Redrawn in part from Ramus (1985).

irrespective of their angle of incidence. There is no unique relationship between recordings from a cosine-corrected and a 4π collector under natural conditions. In a perfectly diffuse radiation field, the response of the sensors should be directly proportional to the surface area of their collectors ($4\pi r^2/\pi r^2$); however, under experimental situations, the ratio may vary from 1 to 4 (Ramus, 1985). Differences in irradiance measured by sensors having the same type of collector have been largely ascribed to calibration errors (Jewson et al., 1984), which are beyond the scope of this discussion.

There are advantages and disadvantages of each collector, and there is no universally ideal sensor for use in phycological research. The type of sensor which should be used depends on the spatial arrangement of the light intercepting surface and the angular distribution of the incident light field. Cosine-corrected sensors were developed mainly for work on terrestrial plant species, where the sensor response largely simulates that of a flat, horizontally oriented surface receiving light from above. Flat, cosine-corrected sensors are most appropriate for studies on some emergent macrophytes and those species in which the photosynthetic structures are found floating on or near the air–water interface or appressed to the substrate. However, many macroalgae do not have rigid, horizontally oriented light intercepting surfaces either in air or water and do not experience unidirectional radiation. Whether to use a 2π or 4π collector under these situations depends on whether the light is unidirectional, that is, received on only one surface, in which case a 2π collector is appropriate, or the light is omnidirectional, that is, received on all surfaces, in which case a 4π collector would be appropriate. For a randomly oriented population of microalgal cells suspended in a diffuse light field, photons are thought to be equally effective for photosynthesis, irrespective of their angular distribution, and a 4π spherical collector would give the most appropriate measure of the incident light level. However, many of the optical parameters used to characterize the underwater light environment experienced by microalgae are based on light fluxes incident on one side of a horizontal surface, for which 2π collectors, which measure radiation equally over a hemisphere, are used.

Unlike field measurements, in which the sensor is suspended above or in the water, it is necessary to distinguish between the amount of light incident *on* the surface of a culture and the light level *within* the culture in laboratory studies. Although the former measurement is often made, the latter measurement is required if laboratory results are to be compared with field investigations. If the light source used in the laboratory is collimated and the incident angle is perpendicular to the flat face of a culture vessel, or the surface of a macrophyte thallus, then the same light level should be found by 2π scalar and cosine-corrected sensors, provided that all other light is excluded. In practice, the light receiving surfaces of the collectors and the culture vessel must be at the same distance from the light source and a flat cosine-corrected sensor may be more appropriate. If, on the other hand, the light source is completely diffuse, the incident irradiance must be integrated over the whole surface of the culture vessel. In this case, the incident irradiance can be determined by averaging measurements made with a 2π sensor at several locations over the exterior of the vessel. A small correction will be necessary for the light attenuated and reflected by the vessel walls. Measurements of irradiance incident *on* the vessel enables the total amount of light absorbed to be calculated, provided that absorptance measurements of the vessel are possible (see below and Osborne and Geider, 1987b). This enables comparison of cultures differing in cell numbers and chlorophyll content directly, without the complications associated with differences in the optical geometries of the measuring and culturing vessel. In theory, a diffuse light source could also be used for the measurements. A sensor immersed within the culture vessel will measure a light level which will be a composite of the absorption and scattering properties of the cells, the optical properties of the culture vessel, and the radiation distribution of the source. Although a spherical (4π) collector would appear to give the best measurement of mean scalar irradiance within the culture, its practical usefulness depends on the relative size of the vessel in relation to the size of the sensor. Light levels can vary severalfold with position inside a vessel, due to directionality and nonhomogeneity of the external light field and the focusing effects of the container wall, so that the appropriate light level within the culture may be difficult to determine.

The sensors described above measure the amount of light available for photosynthesis in the photosynthetically active waveband, but they give no information on its spectral quality. On the basis of the absorption properties of pigments extracted into organic solvents (see Kirk, 1983) the probability of a photon being absorbed should depend markedly on wavelength. In unicellular species and some multicellular plants, absorption is wavelength dependent (Ramus, 1978; Morel and Bricaud, 1981); however, a thick thallus may have absorption characteristics that are largely wavelength independent, presumably as a result of multiple scattering. Spectral measurements of the incident irradiance are required to assess the significance of differences in light quality. Measurements of spectral variations in light quality can be made with a spectroradiometer. Typically, this comprises an energy sensor with a monochrometer interposed between the source and the detector. Commercially available instruments use either irradiance or scalar irradiance collectors and grating or interference monochrometers. New developments now enable instantaneous spectral measurements to be made using diode array detectors (see Chapter 4).

Optical Properties of Cells and Tissues

The absorption and scattering characteristics of algal cells and tissues determines how much incident radiant energy is available for photosynthetic processes. Algal abundance may influence the radiant energy available for photosynthesis, because the characteristics of the underwater light field are in part determined by the absorption and scattering properties of the algae themselves. Light absorption and scattering is also used in a number of experimental techniques for determining the size and composition of algal cells (Yentsch et al., 1983), in remote sensing applications (Smith and Baker, 1978), and for discriminating between phytoplankton species (Wyatt and Jackson, 1989).

Discussions of the optical properties of plants are often complicated because of the wide variety of terms used by workers from different specializations for describing the same optical parameters and because of the misuse of terms such as absorbance and absorptance. The terminology used to characterize the optical properties of algae is summarized in Table 6.3, which conforms to contemporary usage (see Morel and Smith, 1982).

Conventional spectrophotometers generally display results as both absorbance and transmittance. These terms are imprecisely defined, because the optical configurations of spectrophotometers, including the distances between sample compartment and detector and the acceptance angle of the detector, vary widely between manufacturers and models. Transmittance (T) is the decimal fraction of the incident flux transmitted through the sample, which is received by the detector, and absorbance (D) (also referred to as optical density) is defined as the logarithm of the reciprocal of transmittance,

$$T = \Phi/\Phi_0 \tag{6.2}$$

$$D = \log_{10} l/T = \log_{10} T \tag{6.3}$$

where Φ_0 = the incident flux and Φ = the transmitted flux. For a homogeneous pigment solution the absorbance depends on the concentration of the pigment and the pathlength (the Beer-Lambert Law).

$$D = 2.303a^*Cl \tag{6.4}$$

where a^* is the $\log_e$ specific absorption coefficient for a pigment *solution* ($m^2 \cdot g^{-1}$), C is the pigment concentration ($g \cdot m^{-3}$), and l is the pathlength (m). The Beer-Lambert law only holds for pigment solutions and is not applicable to cell suspensions that scatter light. Absorptance is the appropriate measurement of how much light a cell suspension absorbs and should not be confused with absorbance. Although current opinion favors the use of the term absorbance rather than the older expression of optical density, the use of the latter would remove much confusion.

Light attenuation (scattering and absorption) by unicellular plants is

Table 6.3. Summary of terms used to characterize the optical properties of cells.

Term	Symbol	Definition	Units
Absorbance	D	Logarithm to the base 10 of the ratio of the radiant flux incident on a system to the radiant flux transmitted by the system	Dimensionless
Absorptance	α	Fraction of the incident flux absorbed	Dimensionless
Absorption coefficient	a	Quantity of radiant flux per unit pathlength	m^{-1}
Chlorophyll a-specific absorption coefficient	a^*	The absorption coefficient per unit chlorophyll a	$m^{-2} \cdot g^{-1}$ chl a
Absorption cross section	sQ_a	The radiant flux absorbed in a certain cross-sectional area of the incident beam equal to the product of the geometrical cross section (s) and the efficiency factor for absorption (Q_a)	m^2
Attenuation cross section	sQ_c	The radiant flux attenuated (scattered and absorbed) in a certain cross-sectional area of the incident beam equal to the product of the geometrical cross section and the efficiency factor for attenuation (Q_c)	m^2
Attenuation coefficient	c	Attenuation per unit pathlength	m^{-1}
Chlorophyll a-specific attenuation coefficient	c^*	Attenuation coefficient per unit chlorophyll a	$m^2 \cdot g^{-1}$
Backscattering cross section	sQ_{bb}	The radiant flux backscattered in a certain cross sectional area of the incident beam; product of the geometrical cross section and the efficiency factor for backscattering	m^2
Backscattering coefficient	b_b	Quantity of radiant flux backscattered per unit pathlength	m^{-1}
Backscattering ratio	b_b/b	Proportion of total scattered flux that is backscattered	Dimensionless
Chlorophyll a-specific backscattering coefficient	b_b^*	Backscattering coefficient per unit chlorophyll a	$m^2 \cdot g^{-1}$
Efficiency factor for absorption	Q_a	Ratio of radiant flux absorbed by the cell to the flux incident on its geometrical cross section	Dimensionless

Table 6.3. Continued.

Term	Symbol	Definition	Units
Efficiency factor for scattering	Q_b	Ratio of the radiant flux scattered by the cell to the flux incident on its geometrical cross section	Dimensionless
Efficiency factor for attenuation	Q_c	Ratio of the radiant flux attenuated by the cell to the flux incident on its geometrical cross section	Dimensionless
Efficiency factor for backscattering	Q_{bb}	Ratio of the radiant flux backscattered by the cell to the flux incident on its geometrical cross section	Dimensionless
Forward scattering coefficient	b_f	Quantity of radiant flux scattered in the forward direction per unit pathlength	m^{-1}
Forward scattering ratio	b_f/b	Proportion of the total scattered flux that is scattered in the forward direction	Dimensionless
Reflectance	R	Fraction of incident light reflected	Dimensionless
Reflectance ratio	b_b/a	Ratio of the backscattering coefficient divided by the absorption coefficient	Dimensionless
Scattering coefficient	b	Total scattering per unit pathlength	m^{-1}
Chlorophyll a-specific total scattering coefficient	b^*	Total scattering coefficient per unit chlorophyll a	$m^2 \cdot g^{-1}$
Scattering cross section	sQ_b	The radiant flux scattered in a certain cross-sectional area of the incident beam; product of the geometrical cross section and the efficiency factor for scattering	m^2
Transmittance	T	Fraction of the incident light transmitted	Dimensionless
Volume scattering function	$\beta(\Phi)$	Quantitative measure of the angular distribution of scattering radiant intensity in a given direction per unit of irradiance in the cross section of the volume per unit volume	$m^{-1} \cdot steradian^{-1}$
Normalized volume scattering function	$/b=$	Volume scattering function divided by the total scattering coefficient	$steradian^{-1}$

133

dominated by scattering processes (see Table 6.1). Of the total radiant flux incident on a cell, some is absorbed and some is lost by purely physical scattering processes. For a multicellular photosynthetic structure, scattering is normally partitioned into a reflected and a transmitted component and the total flux absorbed (Φ_a) is equal to the incident flux (Φ_0) minus the reflected (Φ_r) and transmitted fluxes (Φ_t),

$$\Phi_a = \Phi_0 - (\Phi_r + \Phi_t) = \Phi_0 - \Phi_s \qquad (6.5)$$

where Φ_s is the scattered flux. The *total* flux absorbed, reflected, or transmitted is extremely difficult to measure. Instead, the fractions of the incident light absorbed (absorptance = α), reflected (reflectance = R), or transmitted (transmittance = T) are determined by

$$\begin{aligned} \alpha &= \Phi_a/\Phi_0 \\ R &= \Phi_r/\Phi_0 \\ T &= \Phi_t/\Phi_0 \end{aligned} \qquad (6.6)$$

Dividing Eq. 6.5 by Φ_0 and noting the identities in Eq. 6.6 leads to the following expressions for absorptance:

$$\alpha = 1 - (R + T) = 1 - \Phi_s/\Phi_0 \qquad (6.7)$$

Arranging the optical configuration of a spectrophotometer to collect all of the scattered light allows the absorptance (α) to be estimated from the absorbance (D), as follows:

$$D = -\log_{10}(1 - \alpha) \qquad (6.8)$$

This equation shows quite clearly the difference between absorbance (D) and absorptance (α). It should be stressed that this relationship can be used to convert absorbance measurements made on a spectrophotometer into true absorption measurements *only if scattering is fully accounted for.* The parameter used extensively for characterizing the absorption of a cellular suspension is the absorption coefficient (a). The absorption coefficient is expressed as an exponential function of the absorptance (α) and the pathlength through the suspension (Kirk, 1983).

$$a = -(1/l) \log_e(1 - \alpha) \qquad (6.9)$$

where a is the absorption coefficient (m^{-1}) and l is the pathlength (m). This relationship is a single scattering approximation and is only valid if multiple scattering does not occur within the suspension (see Morel and Bricaud, 1981). Rearranging Eqs. 6.7 and 6.8 allows the absorption coefficient to be determined from the absorbance

$$a = 2.303 \, D/l \tag{6.10}$$

It is often desirable to know the absorption per unit mass, which is the specific absorption coefficient designated a^*. For example, when expressed per unit mass of chlorophyll a, the chlorophyll a-specific absorption coefficient is

$$a^* = a/C \tag{6.11}$$

where C is the chlorophyll a concentration ($g \cdot m^{-3}$). Rather than normalizing to chlorophyll a, it is sometimes desirable to express absorption on a cell-specific basis. The absorption cross section and the efficiency factor for absorption can be obtained from determinations of the absorption coefficient and measurements of cell dimensions and numbers

$$sQ_a = a/N \tag{6.12}$$

where sQ_a is the absorption cross section ($m^2 \cdot cell^{-1}$), s is the geometric cross section ($m^2 \cdot cell^{-1}$), Q_a is the efficiency factor for absorption (dimensionless), and N is the cell concentration ($cells \cdot m^{-3}$). Substituting $a = a^*C$ into Eq. 6.11 and rearranging leads to

$$Q_a = a^*C/sN \tag{6.13}$$

which upon noting that the chlorophyll a concentration (C) equals the product of the cell concentration (N), cell volume (V, with units in $m^3 \cdot cell^{-1}$), and intracellular chlorophyll a concentration (C_i, with units g chlorophyll $a \cdot m^{-3}$ cell volume) leads to

$$Q_a = a^*VC_i/s \tag{6.14}$$

Expressing the volume and surface area of a cell in terms of an equivalent spherical diameter d by using the geometric formulae $V = \pi d^3/6$ and $s = \pi d^2/4$ allows Eq. 6.13 to be rewritten as

$$Q_a = (2/3)a^*C_id \tag{6.15}$$

The efficiency factor for absorption (Q_a) varies from 0 for a particle that does not absorb light to 1.0 for a highly pigmented particle that absorbs all the light incident on its geometric cross-sectional area. Curvilinearity in a plot of Q_a versus $C_i d$ arises from an inverse relation between a^* and $C_i d$ (Fig. 6.2). The decrease of a^* at large $C_i d$ is due to intracellular self-shading and comprises the phenomena known as the package effect.

Although measuring absorption requires that all scattered photons be collected, measuring attenuation requires that all scattered photons be excluded from the detector. By reducing the collecting half angle to $< 1°$ (Fig. 6.3), scattered photons will be excluded from the detector and only light that has been transmitted through the suspension will be measured. The exclusion of the scattered photons from the collecting angle of the photomultiplier results in a high absorbance reading and a conventional spectrophotometer with this configuration is really measuring attenuated (i.e., scattered and absorbed) light. Similar expressions to those used for determining the absorption coefficients can be used to specify volume and specific attenuation coefficients (c, c^*), the attenuation cross section (sQ_c) and an efficiency factor for attenuation (Q_c) (Eqs. 6.9–6.11). Scattering can be obtained from absorption and attenuation measurements by difference, and comparable parameters can be derived (Table 6.3). The efficiency factor for attenuation (Q_c) is the sum of the absorption and scattering components (Eq. 6.1). Because of the small extent to which microalgal cells backscatter light (see Table 6.1), it is more difficult to measure backscattering than absorption, attenuation, or total scattering. Measurements of backscattering together with total scattering can be used to determine forward scattering coefficients.

Theoretical Treatment of the Optical Properties of Cells and Tissues

Efficiency factors for Q_c, Q_a, and Q_b can be predicted using Mie theory (Mie, 1908), although this involves very complex computations. Instead, the anomalous diffraction approximation of Hulst (1957) is often used. This approximation appears to be appropriate for homogeneous particles or cells with refractive indices up to approximately twice that of the surrounding medium. Expressions for the efficiency factors for attenuation (Q_c) and absorption (Q_a) are given by

$$Q_c = 2.4 \exp{(-\rho\tan\xi)}[\cos\xi \sin{(\rho - \xi)}/\rho \qquad (6.16)$$
$$+ (\cos\xi)^2 \cos{(\rho - 2\xi)}] + 4(\cos\xi)\cos{2\xi}/\rho$$

$$Q_a = 1 + [\exp(-2\tan\xi)(2\tan\xi + 1) - 1]/2\tan^2\xi \qquad (6.17)$$

$$Q_b = Q_c - Q_a \qquad (6.18)$$

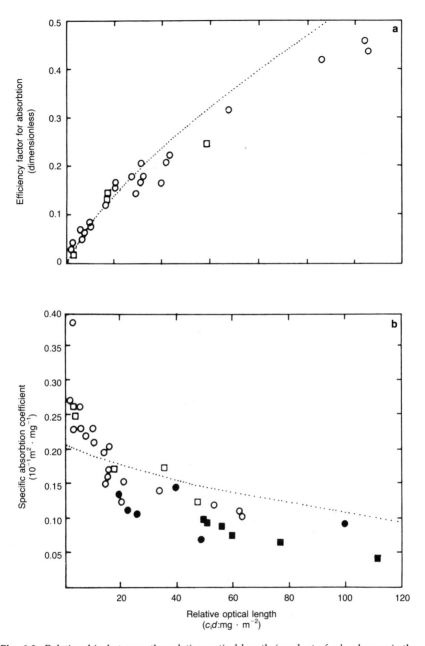

Fig. 6.2. Relationship between the relative optical length (product of $c_i d$, where c_i is the intracellular chlorophyll a concentration and d is the equivalent cell diameter) and the efficiency factor for absorption (a), or the chlorophyll a-specific absorption coefficient (b). Redrawn from the data of (○) Morel and Bricaud (1986), (●, ■) Haardte and Maske (1987), and (□) Osborne and Geider (1989).

137

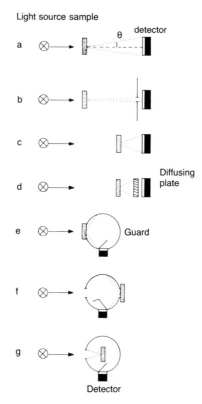

Fig. 6.3. Experimental spectrophotometer configurations for measuring absorption, attenuation (scattering + absorption), and scattering. (a) Conventional spectrophotometer arrangement showing the position of the light source, sample cuvette, and detector. The collecting cone of transmitted light measured by the detector is shown by the dotted lines, and the half angle (θ) is illustrated. (b) Configuration used to measure attenuation, where the collection angle is reduced ($\theta = 0.5°$) so that the detector predominantly measures light that has been transmitted straight through the suspension without encountering a particle. (c) Increasing the amount of light collected by the detector from a scattering suspension by reducing the distance between the sample and the detector or (d) by using a diffusing plate between the sample and the detector. (e) Use of a Taylor-type integrating sphere to capture scattered light, with the sample at the entrance port of the sphere to measure transmittance, or at the exit port (f) to measure back scattering/reflectance, or in the geometric center of an Ulbricht sphere (g) to measure "total" scattering. The detector guard within the integrating spheres ensures that only light that has been reflected off the sphere wall is received by the detector. For further details of the operation and use of integrating spheres and the calculation of optical parameters from spectrophotometer measurements see text.

where $\rho = (4\pi\, d/\lambda)\, (m - 1)$ is the size parameter of the particle with respect to wavelength (λ), diameter (d) and the real part of the refractive index relative to that of the surrounding medium (m) and $\tan\xi = m'/(m - 1)$, where m' is the imaginary part of the refractive index.

The expression for Q_a can be modified so that it is more amenable for many experimental requirements by introducing the parameter $\rho' = 4\pi$ dm'/λ (or $2\rho\tan\xi$). Equation 6.17 then becomes

$$Q_a = 1 + 2\,[\exp(-\rho') + 1]/\rho' + [\exp(-\rho') - 1]/(\rho)^2 \qquad (6.19)$$

The parameter ρ' can be expressed in terms of the absorption coefficient of cell material as follows:

$$\rho' = a_m d \qquad (6.20)$$

where $a_m = 4\pi\,m'/\lambda$ is the absorption coefficient for cell material (m^{-1}) and d is the cell diameter (m). Although these expressions have been used extensively in many recent studies on algae, they *strictly* refer to homogenous spherical particles, and deviations from theoretical predictions may be expected. Also they cannot be applied to situations where multiple scattering occurs.

Theoretical descriptions (Eqs. 6.16–6.19) of the optical characteristics of particles with diameters of 1–20 μm (in the range that is relevant to most microalgal cells) indicate that absorption and scattering is dependent on cell size. Most of the scattering occurs within a small cone in the forward (0°) direction and backscattering (180°) is often less than 1% of the total attenuated light (Kirk, 1983; Osborne and Raven, 1986). Small particles with diameters that are considerably less than the wavelength of light exhibit scattering characteristics with a pronounced wavelength dependence, where scattering is proportional to the fourth power of the wavelength. Although microalgal cells do not approach these dimensions, there is some evidence that very small species ($< 2\,\mu$m diam) may exhibit wavelength-dependent scattering, with greater scattering at short wavelengths (see Osborne and Geider, 1989).

The optical characteristics of suspended cells are called *inherent* optical properties (Preisendorfer, 1961), because their magnitude depends only on the characteristics of the cells themselves. In contrast, vertical attenuation coefficients for light penetration in water bodies are often referred to as *apparent* optical properties, because they are strictly a property of the radiation field. However, the two are often related as the apparent optical properties are largely dependent on the *inherent* optical properties of the aquatic medium (Kirk, 1983).

Spectrophotometric Techniques for Measuring the Optical Properties of Cells and Tissues

The conventional spectrophotometer configuration (Fig. 6.3) shows the relative positions of the light source, sample cell, and detector. In most spectrophotometers, there are two cell positions, one for the reference and the other for sample cuvette, and the light beam is normally split so that it is alternately incident on either cuvette. The detector is normally a photomultiplier, preferably in an end-on configuration. The collecting cone of light received by the detector, defined by the half angle (Fig. 6.3), is small, and the sensor only responds to transmitted light within a few degrees of the forward (0°) direction. Clearly, radiant flux that is scattered at wide (> 10°) angles will not be detected. Erroneously high values for the absorbance in cell suspensions that scatter light can be expected as this instrument only measures the transmitted flux (Φ) that *reaches* the detector. The amount of light that reaches the detector can be increased by increasing the half angle. Conversely, reducing the half angle to $\approx$ 0.2° eliminates most (> 95%) of the scattered light and the detector will only measure the transmitted flux in the forward direction. This approach allows estimation of the scattering characteristics of cell suspensions.

A major problem with studies on the optical properties of plants is to fully account for scattering losses during absorption measurements. Various methods have been utilized to overcome the difficulties of measuring the optical properties of cell suspensions or multicellular tissues using commercial spectrophotometers. Perhaps the simplest way to improve the collection of scattered light is by reducing the distance between the detector and the sample (Privoznik et al., 1978). Alternatively, opal glass (Shibata et al., 1954) diffusers have been used, or the samples have been concentrated on glass fiber filters (Yentsch, 1960, Kiefer and Soo Hoo, 1982; Soo Hoo et al., 1986), so that the filter itself acts as a diffuser, substituting for the opal glass. Both procedures effectively increase the collecting angle of the detector. Radiant flux that is scattered by the sample at wide angles and would normally be lost is collected by the diffuser and reaches the detector (Fig. 6.3). Both the filter and opal glass techniques have, however, been criticized because they are thought to enhance absorption, leading to artificially high absorption coefficients (Kiefer and Soo Hoo, 1982; Maske and Haardt, 1987). Additional absorption enhancement phenomena are also possible because of multiple reflections between particles within the concentrated pigment layer. At high cell concentrations, the absorption per unit pigment may even be reduced in comparison to the equivalent cell suspension (Maske and Haardt, 1987). Samples sandwiched between two filters may exhibit between two- and sixfold increases in absorption (Kiefer and Soo Hoo, 1982; Maske and Haardt, 1987). Although empirical correction factors can be applied to account for these increases (Kiefer and Soo Hoo, 1982; Bannister, 1988; Mitchell and Kiefer, 1988a), there is as yet no quantitative description to enable these techniques to be more applicable. Bannister (1988) has responded to some of the criticisms made by Maske and Hardt

(1987) by showing that predicted errors associated with the use of the opal glass technique are generally small ($\pm$ 10%).

Errors associated with the use of diffusers can be related to inherent differences in the scattering and absorption characteristics of cell suspensions. For instance, particles that scatter predominantly at narrow angles may show little absorption enhancement in comparison to those that scatter at wide angles, because of the reduced possibility of significant increases in total internal reflections within the cuvette at angles of incidence close to the normal (Bannister, 1988). Errors could be influenced by cell size, cell concentration, or pigment content, which argues for more detailed information on the scattering characteristics of cell suspensions. Methodological differences, such as the type of measuring cuvette used (Bannister, 1988) or the relative geometrics of the sample, diffuser, and detector positions may alter the measurements. Difficulties may also arise because the scattering characteristics of the reference cell (generally water) do not match that of the sample cell suspension. Although this may be overcome to some extent by the use of bleached cells (Doucha and Kubin, 1976) or latex spheres of the appropriate dimensions (Ley and Mauzerall, 1982), it remains uncertain to what extent these will match the characteristics of cell suspensions.

Obviously, there are difficulties in the quantitative description of the absorption characteristics of cell suspensions using diffuser techniques. Also, none of these methods account for backscattering or wide ($> 50°$) angle losses, although reflection losses at the walls of the cuvette should be accounted for in dual-beam instruments. An alternative technique, which should theoretically ensure that *all* of the scattered photons are collected, involves the use of an integrating sphere. Typically, this comprises a hollow sphere, although other shapes have been used (Hahne, 1975), coated on the inside with a highly reflective ($> 98\%$, $\lambda = 400$–700 nm) white "paint," generally $BaSO_4$, although formerly spheres were coated with MgO (Grum and Luckey, 1968). $BaSO_4$ suspensions are commercially available (Eastman-Kodak, white reflectance paint), although a 99-to-1, weight/weight mixture of $BaSO_4$ and polyvinyl alcohol can be used (Grum and Luckey, 1968). The white coating ensures that all photons entering the sphere are diffusely reflected with minimal energy losses. A sensor (either a photodiode or a photomultiplier) is used to measure the irradiance within the sphere. A baffle covers the sensor, so that only diffuse light reflected from the wall of the sphere is measured.

Basically, there are two types of sphere, the Taylor-type (Taylor, 1920) and the Ulbricht sphere (Ulbricht, 1920; Rabideau et al., 1946). In the Taylor-type sphere, reflectance (i.e., backscattering by algal suspensions) and transmittance are measured as separate components of the total attenuated light (Fig. 6.3). With the sample at the front port of the sphere, only transmitted photons are measured, whereas reflected (backscattered) photons are measured with the sample at the rear port. The measurements are calibrated with reference to the appropriate absorption/reflectance standards. Even though transmittance and reflectance measurements are used for characterizing the optical properties of macroalgae, reflectance (backscattering) measurements are generally omitted from determinations made on microalgal suspensions. Backscattering by algal suspensions is difficult to measure because reflection from the surface of the measuring cuvette may exceed backscatter by the algal suspension. Conventionally, measurements involving microalgae are recorded in the absorbance mode, whereas measurements on macrophytes are recorded in the transmittance mode. For macrophytes, absorptance determinations are reported directly as the fraction of the incident light absorbed, whereas absorption by algal suspensions is usually quoted in terms of the absorption coefficient (a).

In the Ulbricht sphere, the sample is placed in the geometric center (Fig. 6.3). With this configuration, the absorptance can be measured directly and theoretically *all* scattering losses can be accounted for. As far as we are aware, this type of instrument is not commercially available and custom-built models allow the measurement of absorptance and scattering, not absorbance. With the sample in the beam, the sensor receives a fixed proportion of the light that is scattered from the sample and a modified form of Eq. 6.7 can be used with a small correction for the component of the diffuse radiation absorbed by the sample while it is inside the sphere. Absorptance can be calculated from

$$\alpha = 1 - \frac{\text{detector response with the sample in the beam}}{\text{detector response with sample out of the beam}} \quad (6.21)$$

where the denominator corrects for the absorption of diffuse radiation within the sphere. Absorption coefficients can be calculated from Eq. 6.9. In recent developments of the Ulbricht sphere for use with terrestrial plants, the sample has been illuminated with diffuse light (Idle and Procter, 1985; Oquist et al., 1978), although there are no published accounts of such measurements being made on algae.

Although the integrating sphere should account for all scattered light, potential operating errors caused by often unavoidable deficiencies in sphere construction may be encountered. There is little information on the use and construction of integrating spheres or of the factors that influence their performance in the phycological literature. Because of the importance of accuracy in optical measurements, particularly for determinations of photosynthetic efficiency, integrating spheres are considered here in more detail. Full descriptions of the performance of various types of integrating spheres are given in the following references (Ulbricht, 1920; Taylor, 1920; Latimer, 1956; Jacquez and Kuppenheim, 1955; Reule, 1962; Birkebak and Cho, 1967; Safwat, 1970). We assume that errors associated with the measurement of the *actual* flux attenuated by the sample are minimal. These may arise if the measuring beam has a larger area than the sample or if back reflection causes losses within a few degrees of the incoming flux (Latimer, 1942). This discussion is restricted to aspects of sphere construction and performance that influence the accuracy of reflectance measurements.

Considering first a sphere with no internal mountings, the detector response will depend on the total flux (mol·s^{-1}) incident on the sphere wall (Φ_w). Φ_w is the product of the incoming flux (Φ_i), the reflectance or transmittance of the sample or standard, depending on the configuration used (see above), and the efficiency with which the sphere retains the incoming flux, often called the throughput (t), Φ_w/Φ_i. The throughput takes into account errors associated with sphere construction, including the reflectance of the inner wall. For reflectance measurements,

$$\Phi_w^{st} = \Phi_i R_{st}\, t_{st} \tag{6.22}$$

and

$$\Phi_w^{s} = \Phi_i R_s t_s \tag{6.23}$$

where R is the reflectance and the symbols s and st refer to the sample and standard, respectively. Optimization of sphere design is most easily carried out in terms of reflectance, rather than scattering, as reflectance standards are readily available. If the measurements are carried out by substitution of the sample for the standard, $t_s \neq t_{st}$. This is due to differences in reflectance between the sample and the standard and because the reflectance of all portions of the sphere wall will not be the same. This error increases as the reflectance of the sample deviates increasingly from that of the standard. The error, associated with this can be obtained from

$$t_s/t_{st} = 1 + x \tag{6.24}$$

The throughput (t_s/t_{st}) can be calculated using equation 6.25. Direct comparison of the samples and standards, for instance, in dual beam spectrophotometers will result in lower errors, although these may not be insignificant as the samples/standards used are flat, so that the effective sphere surface deviates somewhat from perfect sphericity (Jacquez and Kuppenheim, 1955). The efficiency of the sphere (throughput) is a function of the reflectance of the sphere coating and sphere geometry, particularly the number and size of ports in the sphere wall. For a particular sphere the efficiency or throughput, can be determined from the following:

$$t = f_e R_c / [1 - R_c(1 - f_t)] \qquad (6.25)$$

where f_e is the area of the entrance port divided by the total surface area of the sphere, f_t is the total area of all ports divided by the surface area of the sphere, and R_c is the reflectance of the sphere coating.

Losses can be significant in large spheres because many reflections are required before the flux reaches the detector. Small spheres are more efficient and provide the highest throughput, although this will be associated with an increase in the error (x), as the percentage of port area to internal surface area will be higher. Consequently, in the design of an integrating sphere, the efficiency should be as large as possible and the error should be small (maximizing t and minimizing x). To do this, the reflectance of the internal coating should be as high as possible and the ratio of the size of the sample and entrance ports relative to the total surface area of the sphere should be as small as possible. In contrast, detector port size has little effect on efficiency. Therefore, the appropriate sphere size represents a compromise between a requirement for a large sphere to minimize x and a small sphere to maximize t.

Another important consideration is the positioning and type (photodiode or photomultiplier) of detector to be used. For instance, the detector response should be the same irrespective of the angular distribution of the flux within the sphere. A convenient test uses varying concentrations of a scattering suspension in the sample compartment. If only one detector is used, it should be positioned close to the entrance port in the upper portion of the sphere (Latimer, 1956). However, positioning may vary due to differences in the scattering characteristics of the sample and the sphere configuration used (Birkebak and Cho, 1967; Safwat, 1970). Alternatively, a blank with an angular dependence of scattering that is similar to that of the sample could be used, although this requires more detailed information on the scattering characteristics of algal cells and tissues. Increasing the size of the detector could improve the measurements but has obvious limitations, because a flat (single) detector will decrease the sphericity of the sphere. To overcome this problem several small sensors could be positioned at various points on the sphere wall, although this has not been used extensively.

The choice of detector depends on the value of Φ_w. With a low value for Φ_w, a photomultiplier may be required, whereas a photodiode may be acceptable if the value of Φ_w is high. The flux reaching the sphere wall can be calculated from geometric considerations using the sphere efficiency or throughput, from the following expression:

$$\Phi_w = \Phi_i R[tA_d \, (d_e^{2/4} + D^2)] \tag{6.26}$$

where A_d is the area of the detector, d_e is the entrance port diameter, and D is the distance from the entrance port to the detector. The range of expected fluxes can be determined using either R_{st} (maximum) or R_s (minimum).

Additional complications in the use of integrating spheres may arise when the sample is placed within the sphere or because of the use of baffles to shield the detector. This results in blockage effects and the formation of shadows or areas of reduced reflectance on the sphere wall. Absorption/reflection of the internal flux by the sample can be corrected by recording the detector response after moving the sample to a position to one side of the incoming flux. The internally located sample also alters the radiation distribution within the sphere, resulting in areas of high reflectance on the sphere wall. The influence of the sample on the internal flux depends on sample size and the reflecting and absorbing properties of the sample. Generally, these effects can be minimized if the sphere diameter is at least 1.5 times the largest dimension of any device mounted or located within the sphere.

Commercially available integrating spheres are generally of the Taylor type, in which the samples are located externally, so that blockage effects should be minimal. However, samples for use with Taylor spheres should be abutted as closely as possible to the inner wall and preferably the ports should have knife-edge openings, in order to maximize the collection of scattered light at wide angles. Even if these recommendations are followed, the radiant flux scattered at extremely wide angles ($\approx$ 100–180°) will not be captured in the Taylor sphere. The significance of losses at wide angles will depend on the volume-scattering function of the material used. Scattering losses through the side walls of cuvettes can be reduced, if narrow pathlength vessels are used. Notwithstanding these reservations, it would seem that the integrating sphere is the most appropriate instrument for measuring the optical properties of algae in the absence of complete (4π) angular measurements of scattering. Integrating spheres can now be custom built to suit particular requirements (e.g., Oriel Scientific Ltd.). The Ulbricht sphere, in particular, seems ideally suited for measurements on microalgal suspensions, because it accounts for all scattering losses, even those at very wide angles. This sphere configuration has the added advantage that total scattering determinations, including the back-scattered flux, can be conveniently and more accurately made without the requirement for separate measurements of absorption and scattering.

Total scattering measurements in microalgal suspensions are normally determined from separate measurements of absorption (ignoring back scattering) and attenuation. Conventional spectrophotometers can be modified to measure attenuated light by abutting narrow (<< 1 mm) knife-edge slits against the front of the photomultiplier housing, with slit widths chosen to reduce the collecting half angle to < 0.2°. For most cells, this should exclude more than 95% of the scattered light, so that the attenuation measurements will be underestimated by less than 5% (see Bricaud et al., 1983). Further reduction in the slit widths and collecting angle is not practical.

The components of the scattered flux (reflectance and transmittance) within multicellular tissues can be determined with either a Taylor or Ulbricht sphere. With the Taylor sphere, the samples are located in turn at the forward and rear ports for measurements of transmittance and reflectance, respectively. Alternatively, in the Ulbricht sphere, total scattering measurements are made initially and then reflectance determinations are made by placing a highly absorbing backing behind the sample, so that the transmitted flux is eliminated and the flux measured in the sphere is due to only the reflectance component. Transmittance is determined by difference. In theory, similar techniques involving the use of either integrating sphere are appropriate for the measurement of the back-scattered flux from unicellular suspensions. In practice, these measurements are difficult because of the low back-scattered signal obtained with most algal cultures (see Bricaud et al., 1983). With the Taylor sphere, there is also the problem of losses through the side walls of the cuvette.

Angular measurements provide additional information on the optical characteristics of cells. The angular distribution of the scattered flux can be determined at regular intervals and integrated over π to 2π (back scattered), $-\pi$ to π (forward scattered) and 0 to 4π (total scattered) radians, using a rotating detector on a calibrated turntable (see Aughey and Baum, 1954; Spinrad et al., 1978). These measurements are rarely made, largely because the forward-scattered flux predominates and is of an extremely high radiance in comparison to scattering at other angles. First, it is physically very difficult to measure scattering accurately at narrow angles within a few degrees of the forward direction, and, second, the detector response has to be accurate over six-orders-of-magnitude variation in radiance (see Morel and Bricaud, 1986). New techniques using diffraction gratings (Philipona, 1987) and optical fibers (Brown, 1987) in conjunction with photon-counting detectors, should give more accurate estimates of narrow-angle scattering by unicellular algae.

Absorption and scattering measurements made on cell suspensions provide optical parameters based on an average cell or group of cells in that population; they give no information on the optical characteristics of individual cells or of the variation within that population. Recently, a microspectrophotometric technique was described for measuring the efficiency factors for absorption by individual particles or cells (Iturriaga et al., 1988), allowing the identification of the major components that contribute to light attenuation in natural samples. Problems with this technique include (1) the limited depth of field associated with microscope optical systems, which results in the illumination of only the portion of the sample within the plane of focus, (2) a tendency for cells in sample preparations to align along a preferred orientation, which may bias absorption characteristics, and (3) uncertainty in the correction for scattering. Localized light-scattering measurements of individual phytoplankton cells has recently been described (Wyatt and Jackson, 1989), and these instruments are likely to reveal more information on the optical characteristics of unicellular species.

Assessment of the Package Effect

A consequence of the localization of light-absorbing pigments within cells or organelles is to reduce their absorption per unit pigment in comparison to the same quantity of pigment dispersed in solution (Fig. 6.4). This effect is referred to as the *package effect* (also known as the *sieve effect*) and results in a reduction in absorption at the absorption maxima with little change at those wavelengths that are only weakly absorbed (Duysens, 1956; Kirk, 1983). For unicellular algae, the magnitude of the package effect depends on cell size and intracellular pigment content (Morel and Bricaud, 1981). Large cells, or those with a high pigment content, have a greater package effect than small cells or those with a lower pigment content. Changes in the magnitude of the package effect often produce variations in the chlorophyll a-specific absorption coefficient (a^*), which is a measure of the amount of light absorbed per unit chlorophyll a. Although the presence of accessory pigments can modify the value of a^* (Berner et al., 1989), low values are generally indicative of a large package effect and high values are associated with a small package effect (Morel and Bricaud, 1981).

The general features of the package effect in unicellular species have been identified, and experimental observations have largely confirmed theoretical predictions. However, the experimental assessment of the magnitude of the package effect is not without difficulties in the estimation of scattering losses (Osborne and Geider, 1989). There is as yet no

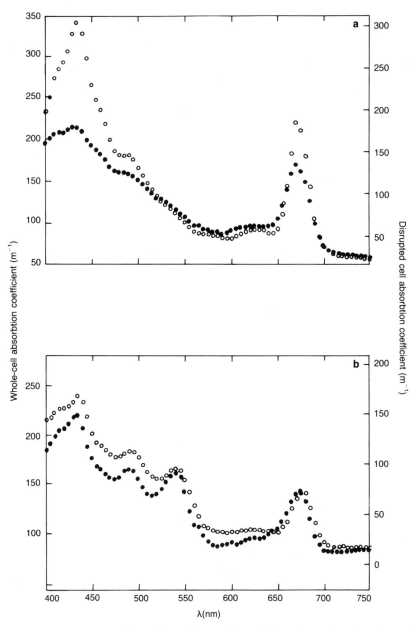

Fig. 6.4. Demonstration of the package effect in a large *Thalassiosira* species (a) and a small *Synechococcus* species (b), using measurements of intact (●) and disrupted (○) cell absorption spectra. Note the increase in absorption after cell disruption is more pronounced in the larger ≈5-μm-diam *Thalassiosira* sp. than in the smaller ≈1.5-μm-diam *Synechococcus* sp. (Osborne and Geider, 1989).

method to quantify the package effect in multicellular tissues, although this has been estimated using Kubelka-Munk theory (Fukshansky, 1981). Indirect evidence indicates that the flattening of the absorption spectrum in multicellular tissues is due to enhanced absorption of wavelengths that are weakly absorbed and is not due to the package effect (Osborne and Raven, 1986), but this requires further investigation.

There may be benefits associated with the enclosure of light-absorbing pigments within organelles (Osborne and Raven, 1986). There is now good evidence that variations in the magnitude of the package effect in thin multicellular tissues resulting from changes in chloroplast orientation (see Britz and Briggs, 1987) can have a considerable effect on light absorption and may reduce the possibility of photoinhibitory damage (Osborne and Raven, 1986). Chloroplast aggregation in response to high light has also been observed in some unicellular species (Kiefer, 1973), where it may serve a similar function. Chloroplast movements and orientation changes are also common in more complex multicellular tissues (Zurzycki, 1961), where they may modify light gradients through photosynthetic tissue.

The first estimates of the package effect in unicellular species used Mie theory to calculate the in vitro absorption spectra from in vivo measurements (Morel and Bricaud, 1981) using Eqs. 6.16–6.18. However, uncertainty in predicting the optical characteristics of cells based on theoretical models that do not account for inhomogeneities in intracellular pigment distribution argues for an experimental approach to this problem. A technique is required to remove quantitatively all of the pigment without altering its in vivo absorption characteristics. Extraction of the light-absorbing pigment by organic solvents may disrupt macromolecular interactions between chlorophyll and other organic molecules and will alter in vivo absorption characteristics, making organic solvents unsuitable extractants for estimating the package effect.

One technique (Fig. 6.4) to assess the package effect involves the disruption of intact cells by sonication or high-pressure treatments at low temperatures (Das et al., 1967; Geider and Osborne, 1987; Osborne and Geider, 1989). These treatments break down cell structures, leaving predominantly thylakoid fragments (Graham and Smillie, 1971). Cell disruption by high pressure in an X-Press (L.K.B. Biotec) proved particularly useful with recalcitrant cells that were not easily fragmented by sonication and has the added advantage of being carried out at subzero temperatures ($-40°C$) at which pigment breakdown should be minimal. However, good agreement was found between the two techniques with cells that were more easily disrupted (Geider and Osborne, 1987). Pigment breakdown can be prevented by sonicating cells under an inert atmosphere (Das et al., 1987). Another technique involves solubilization of thylakoid membranes using the nonionic detergent Triton X-100 (Berner et al., 1989). This technique does, however, produce spectral shifts in absorption in the red region of the spectrum. However, these shifts appear to be insensitive to the concentration of detergent used, and the characteristic shape of the in vivo absorption spectra was retained in the blue region (Berner et al., 1989).

Although these techniques can give an experimental assessment of the overall significance of the package effect, they cannot resolve the separate influences of cell size, chloroplast size or numbers, or variations in thylakoid structure. Indirect estimates by Berner et al. (1989) suggest that the optical properties of the thylakoids have the greatest influence on the magnitude of the package effect. However, quantification of the package effect using these techniques is presently hampered by an inability to ascribe attenuation measurements at $\lambda = 750$ nm to either scattering or absorption (see Osborne and Geider, 1989). Conventionally, the attenuation at $\lambda = 750$ nm is ascribed to scattering, and a flat baseline correction is applied to the absorption spectra between 400 and 700 nm. However, this apparent absorption is found even in optical configurations where all of the scattered light should, on the basis of theoretical predictions, have been collected (Morel and Bricaud, 1981; Bricaud et al., 1988; Osborne and Geider, 1989). Further improvements in the techniques used for assessing the magnitude of the package effect are needed. In addition, more detailed experimental information on the absorption and scattering properties of algal cells and tissues is required. As yet, there are no experimental techniques for measuring the package effect in multicellular tissues and present theoretical treatments (Fukshansky, 1978) have too many constraints to be of general practical use.

Theory and Measurement of Action Spectra

An effective means of utilizing the *available* light would seem to be a prerequisite for survival in the aquatic environment. Many species have to contend with considerable spectral variations in light quality on time scales of minutes to days due to water movements, the extent of submergence and/or because of variations in the optical properties of water. Although the distribution of many plants is often correlated with marked differences in light quality, a causative explanation for the distribution of algae based on differences in light quality is currently contentious (Dring, 1981; Ramus, 1982). Evaluation of the significance of spectral variations in light quality on photosynthetic performance and the effectiveness of different combinations of accessory pigments can be assessed by measurements of action spectra. These are determined from measurements of photosynthetic rate at limiting light levels over narrowly defined spectral bands (25 nm or less) in the range 300–750 nm. Under these conditions, the rate of photosynthesis (P) at a particular wavelength will be determined by the product of the photon yield on an absorbed light basis (ϕ_m), the incident irradiance (E), and the absorptance (α), which is a measure of the amount of light absorbed,

$$P(\lambda) = \phi_m (\lambda)E(\lambda) \, \alpha(\lambda) \qquad (6.26)$$

If the photon yield, ϕ_m, is invariant, the action spectrum should have the same shape as the absorptance spectrum. Differences between the action spectrum and the absorption spectrum will reflect differences in the ability of individual pigments to utilize the absorbed light in photosynthetic reactions. Accessory pigments that exhibit little or no transfer of excitation energy to the chlorophyll a pigment-protein complexes will have a low $\phi_m(\lambda)$, and this will result in a low photosynthetic rate. With macrophyte tissues, the photosynthetic rate is generally expressed on a unit area basis and P is expressed in units of micromoles of O_2 per square meter per second. As E is expressed in micromoles of photons per square meter per second and α is dimensionless, ϕ_m is expressed in moles of O_2 per mol photon. With microalgal suspensions, P is generally expressed on a unit chlorophyll (or biomass) basis and the rate of photosynthesis at any given wavelength is given by

$$P^{chl}(\lambda) = \phi_m(\lambda)E(\lambda)a^*(\lambda) \qquad (6.27)$$

where a^* is the in vivo chlorophyll a-specific absorption coefficient. A better way to examine variations in the value of $\phi_m(\lambda)$ attributable to various accessory pigments is to plot the photon yield against wavelength. In the past, single measurements of photosynthesis and absorbed light [$E(\lambda)\,\alpha(\lambda)$] were used to estimate ϕ_m. In view of the difficulty of making photosynthesis measurements at limiting light levels, ϕ_m should strictly be determined from the slope of $P(\lambda)$ versus $E(\lambda)\,\alpha(\lambda)$. Measurements of ϕ_m should also be made at values of E above the light compensation point, where transitory respiration phenomena (such as those associated with the Kok effect) are absent (Kok, 1949; Osborne and Geider, 1987a, b). The irradiances employed by some investigators may not always be above the light compensation point, particularly in the blue part of the spectrum (e.g., Dring and Lüning, 1985; Lewis et al., 1985) and may not give an accurate assessment of the role of individual pigments in light utilization. Xenon arc sources (Chapter 4) are now widely available and give the necessary output and light quality required for action spectra determinations at short wavelengths (Halldal, 1964; Scott and Jitts, 1977).

The measurement of photosynthetic action spectra requires the determination of photosynthesis rates at discrete wavelengths using O_2, CO_2, or ^{14}C exchange. In the past, manometric techniques (Emerson and Lewis, 1943) were used to determine action spectra based on net O_2 exchange, but these have been superseded by O_2 electrodes. Action spectra for net CO_2 exchange have, as for as we know, been restricted to measurements made on terrestrial plants. Although the ^{14}C technique has been used in laboratory studies (Iverson and Curl, 1973), its usefulness is particularly evident for "field" measurements of action spectra on dilute algal suspensions where O_2-exchange techniques are of insufficient precision (Lewis et al., 1985). Isolation of the spectral bandwidths can now be achieved with a wide range of commercially available narrow-bandpass interference filters (see Dring and Lüning, 1985; Lewis et al., 1985), although high-intensity monochromators (Haxo and Blinks, 1950; Scott and Jitts, 1977; Neori et al., 1986; 1988) are to be preferred.

In general, the action spectra of multicellular green and brown algae are similar to those reported for a number of terrestrial vascular plants (Fig. 6.5), although measurements on green unicellular species exhibit a lower absorption in the green region of the spectrum (see McCree, 1972; Inada, 1976; Lüning and Dring, 1985; Neori et al., 1986). In contrast to the green and brown algae, the action spectrum of the red algae shows a marked peak around 500–600 nm, with little photosynthesis in either the blue or red ends of the spectrum (Haxo and Blinks, 1950; Lüning and Dring, 1985; Neori et al., 1986, 1988). Cryptomonads exhibit similar responses to those shown for the red algae but may have higher photosynthesis rates in the blue and red spectral regions (Neori et al., 1986, 1988). It is also evident that near-ultraviolet wavelengths should not be ignored in assessing the photosynthetic performance of algae (see Halldal, 1964; McLeod and Kanwisher, 1962; McCree, 1972; Inada, 1976).

Superficial agreement is usually observed between action and absorptance spectra in species of the Chlorophyta, Bacilliarophyta, and Pyrrophyta (Haxo and Blinks, 1950; Lüning and Dring, 1985; Lewis et al., 1987; Neori et al., 1986, 1988). Apart from the data of Lewis et al. (1987), all these species show high absorption and photosynthesis in the blue region of the spectrum, and this discrepancy may be related to methodological problems (see above). However, in phycobili protein-containing phytoplankton (Rhodophyta, Cryptophyta, and Cyanobacteria), the action spectrum largely reflects light that is absorbed by the bilipigments (Phycoerythrin and Phycocyanin) in the 500–600 nm waveband, and low photosynthesis rates are found in the blue and red spectral regions (Haxo and Blinks, 1950; Luning and Dring, 1985; Lewis et al., 1986, 1988; Neori

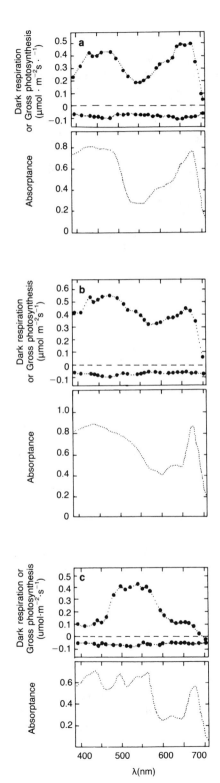

Fig. 6.5. Photosynthetic action and absorption spectra of the macroalgae *Ulva lactuca* (a), *Laminaria saccharina* (b), and *Delesseria sanguinea* (c). Redrawn from the results of Luning and Dring (1985).

et al., 1986, 1988). Current opinion largely favors the interpretation that spectral variations in ϕ_m are a result of alterations in the distribution of excitation energy between PS1 and PS2, and that optimal photosynthetic performance can only be achieved when these are balanced (see Evans, 1987; Neori et al., 1986). The well-known red-drop, for instance, at wavelengths > 680 nm is ascribed to a deficiency in light absorbed by PS2. Although the drop in the blue is often related to absorption by nonphotosynthetic cell components (McLeod and Kanwisher, 1962) or photoinduced damage (Halldal, 1964), this could also be due to a reduction in the amount of light absorbed by PS2. Bilipigments transfer excitation energy exclusively to PS2, whereas most of the chlorophyll in phycobilin-containing species is associated with PS1 (Larkhum and Barrett, 1983). Light absorbed by chlorophyll in the blue and red spectral regions is largely ineffective in photosynthesis due to overexcitation of PS1 (Larkhum & Barrett, 1983), resulting in a mismatch between the absorptance and action spectra. Evidence from laboratory (Wyman et al., 1985; Neori et al., 1986) and field measurements (Lewis et al., 1987) suggest inefficiencies in energy transfer to PS2 by phycobiliproteins under some conditions. However, ascribing differences in absorptance and action spectra based solely on balanced excitation of PS2 and PS1 ignores internal controls that can regulate the distribution of energy between the two photosystems via State 1–State 2 transitions or spillover (Fork and Satoh, 1986).

As energy transfer can only occur from PS2 to PS1, all action spectra that are determined without supplementary light may be biased toward the absorption spectrum of PS2 and may be a poor guide to the ability of a plant to utilize the mixture of wavelengths that are available under natural conditions (Kirk, 1983). It is evident that the addition of light, predominantly absorbed by PS2, can result in the enhancement of photosynthesis in the blue (Fork, 1963) or red (Emerson et al., 1957) regions of the spectrum, indicating that many or all of the accessory pigments are photosynthetically competent. Action spectra may be used, not only for examining which pigments transfer light energy to chlorophyll a (in which case supplementary light should be used) but also for measuring the ability of a species to photosynthesize in light of different spectral qualities (in which case they should be given in isolation) (Lüning and Dring, 1985). Although some enhancement is likely to occur under natural conditions, it is uncertain whether the available wavelengths would result in the balanced operation of PS1 and PS2. Measured and calculated responses of macrophytes to representative broad-band underwater light fields provides no evidence for enhancement under commonly occurring spectral regimes (Dring and Lüning, 1985), consistent with data for terrestrial vascular plants. As a general protocol, however, action spectra should be measured both in the presence and in the absence of supplementary light.

As an alternative to O_2 evolution (or CO_2-exchange), in vivo fluorescence excitation spectra may be more conveniently and readily measurable, particularly under field conditions (Yentsch and Yentsch, 1979; Mitchell and Kiefer, 1984). If the fraction of open-to-total PS2 reaction centers is maintained at a constant level, fluorescence and O_2-evolution spectra should, theoretically, have similar characteristics (Neori et al., 1986). Recent studies have demonstrated good agreement between fluorescence and O_2 measurements, using a modulated fluorescence technique and a strong background light to maintain the fraction of open-to-total PS2 reaction centers constant (Neori et al., 1986; see also Chapter 3). There may, however, be problems in using this technique with populations dominated by bilipigment-containing organisms, where PS2 fluorescence measurements may underestimate photosynthetic performance. Also, there are problems in extending this technique to measurements of natural phytoplankton populations (discussed by Neori et al., 1986), which require further examination.

Chapter 7

The Photosynthesis–Light Response Curve

Instantaneous rates of photosynthesis may be controlled by external factors, including irradiance, temperature, and the concentrations of CO_2 and O_2, by the molecular composition of the photosynthetic apparatus, including the concentrations or activities of pigments, catalysts of electron transport, and enzymes of CO_2 fixation, or by other factors, such as the concentration of metabolites, or the demands for photosynthetic products. Photosynthesis-factor response curves, such as the photosynthesis–irradiance, photosynthesis–CO_2, and photosynthesis–temperature relations describe the response of photosynthesis rates to external factors. In addition, the interpretation of these curves often provides important information on regulatory control of photosynthesis by the composition of the photosynthetic apparatus and other internal factors.

Interpretation of resource–response relations rests on the basic principle formulated by Blackman (1905) that an abrupt break in the response curve indicates a switch in control of photosynthesis between two different external factors. Often, these external factors have their dominant influence on different components of the photosynthetic apparatus. For example, the efficiency of utilization of irradiance often limits the initial slope of the photosynthesis–irradiance (PI) curve, but temperature or CO_2 availability often determines the light-saturated rate. In other words, the initial slope is often determined by light harvesting and the maximum rate by carboxylation. Usually, the simple resource–response relation envisaged by Blackman or a form based on Michaelis-Menten kinetics is not observed, in part because the concentrations of CO_2 and O_2 or irradiance at the site of photosynthesis within a microalgal cell or macroalgal thallus differs from that in the external environment. In an attempt to deal with this problem, plant physiologists have developed models based on a mechanistic description of transport and fixation of CO_2 and irradiance distribution within leaves (Gutschick, 1984). Although gradients in irradiance and resistances to CO_2 diffusion may be particularly important in macroalgae (Chapter 6), models similar to those employed with terrestrial vascular plants have not been applied to the algae.

The response of photosynthesis to irradiance can be considered over the short and long term. Instantaneous observations of CO_2 or O_2 exchange made over the short term are often considered to be uncoupled from growth, whereas the cumulative rate of photosynthesis should be consistent with the measured growth rate. In practice, however, it is often difficult to separate photosynthesis from growth. Many techniques require finite time intervals for carrying out photosynthesis measurements during which growth can occur. Although it is often taken for granted that algal growth rate can be predicted from measurements of photosynthesis rate, it was not until 1979 that Bannister demonstrated that measurements of growth and photosynthesis in the green algae *Chlorella pyrenoidosa* were internally consistent. The difficulty in relating growth to photosynthesis arises largely from phenotypic, genotypic, and ontogenetic changes in the size and composition of the photosynthetic apparatus with consequent changes in biomass-specific photosynthesis rates. Interpretation of the response of algal growth to irradiance has been given a mechanistic foundation in terms of the energetic efficiency and size of the photosynthetic apparatus (Bannister, 1979; Shuter, 1979; Kiefer and Mitchell, 1983; Geider, 1990).

The PI curve plays a central role in measuring, modeling, and predicting algal photosynthesis, and in assessing both intra- and interspecific variations in photosynthetic physiology. It contains important information on the functioning of various components of the photosynthetic apparatus and their response to environmental variables. This chapter considers the equations that are used to describe the PI curve, the range of variability of the parameters of the PI curve, and problems encountered in measuring the PI curve.

Mathematical Descriptions of The Photosynthesis–Irradiance Curve

The relationship between net gas exchange and irradiance can be broken up into three regions; these are a light-limited region in which photosynthesis increases with increasing irradiance, a light-saturated region in which photosynthesis is independent of irradiance and a photoinhibited region in which photosynthesis decreases with further increases in irradi-

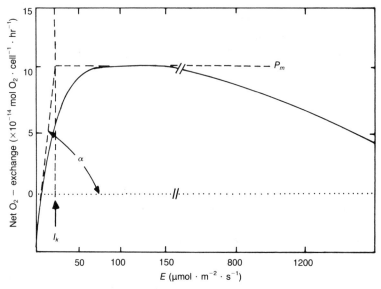

Fig. 7.1. Influence of irradiance on the photosynthetic rate of *Chlamydomonas rheinhardtii* previously grown at 15 μmol·m^{-2}·s^{-1}. P_m is the maximum rate of light saturated photosynthesis, α is the initial slope, and I_k ($I_k = P_m/\alpha$) is the irradiance at the onset of light saturation. Redrawn from Neale (1987).

ance (Fig. 7.1). In addition some workers have identified an intermediate region, where there is either a gradual or abrupt transition from light-limited to light-saturated photosynthesis (see Walker, 1989; Sharkey, 1989). The photosynthesis-light response (PI) curve is constructed by plotting the biomass-specific rate of net photosynthesis against irradiance. Biomass may be expressed in terms of a number of possible variables including chlorophyll *a*, total pigment content, organic carbon, organic nitrogen, or dry weight.

Mathematical descriptions or models of the PI curve are empirical in that they attempt to describe the influences of all of the factors that may affect photosynthesis rates in terms of the minimum number of parameters required to adequately describe the PI curve. Most models of the PI curve characterize the light-limited and light-saturated regions with three parameters: these are the initial slope (α), the dark respiration rate (R), taken as the extrapolation of the initial slope to darkness, and the light-saturated photosynthesis rate (P_m). Inclusion of the photoinhibited region requires additional coefficients in the mathematical formulation and complicates the interpretation of the meaning of the parameters. This section begins with a consideration of light-limited and light-saturated photosynthesis and then proceeds to an examination of the complete PI curve, including photoinhibition.

Table 7.1.   Different formulations of the photosynthesis–light response curve for gross photosynthesis in the absence of photoinhibition.

$P = \alpha E$	$E \leq P_m/\alpha$	(7.1)
$P = P_m$	$E > P_m/\alpha$	
$P = \dfrac{P_m \alpha E}{P_m + \alpha E}$		(7.2)
$P = \dfrac{P_m \alpha E}{[P_m^2 + (\alpha E)^2]^{0.5}}$		(7.3)
$P = \alpha E \exp(\alpha E/P_m e)$		(7.4)
$P = \alpha E \exp(-\alpha E/P_m e)$	$E \leq P_m e/\alpha$	
$P = P_m$	$E > P_m e/\alpha$	(7.5)
$P = P_m [1 - \exp(\alpha E/P_m)]$		(7.6)
$P = \alpha E - [\alpha E - (\alpha E)^2]/4P_m$	$E \leq 2P_m/\alpha$	(7.7)
$P = P_m$	$E > 2P_m/\alpha$	
$P = P_m \tanh(\alpha E/P_m)$		(7.8)

Modified from Jassby and Plat (1976).

The quantitative description of the light-dependence of photosynthesis dates from Blackman's (1905) studies of limiting factors in plant production. Blackman considered the rate of plant production to be linearly dependent on the availability of a single limiting factor at low resource supply and independent of the availability of this factor above some threshold value (Eq. 7.1, Table 7.1). It was soon recognized that the transition between limiting and saturating resource availability may not be as abrupt as postulated by Blackman kinetics, and a number of different formulations of the PI curve were proposed by plant physiologists, oceanographers, and limnologists (Table 7.1). Although algal physiologists often take the shape of the PI curve as fixed, it may in fact be variable, and vascular plant physiologists often explicitly employ a shape factor (Leverenz, 1987). Abrupt transitions in the PI curve are thought to result from extremely efficient regulatory controls over the activity of the primary carboxylating enzyme RUBISCO (Sharkey, 1989), although these may be obscured due to gradients in the internal light environment. Variations in the relative importance of these biochemical and nonbiochemical factors could have important consequences for mathematical descriptions of photosynthetic responses to irradiance as the greatest imprecision in current models is often found in the transition region.

In analogy with the Michaelis-Menton description of enzyme kinetics, the rectangular hyperbola (Eq. 7.2, Table 7.1) has been used to simulate the photosynthesis–light response in ecological models. Unfortunately, this equation consistently yields a poor fit to experimental data, especially in the transition region from light limitation to light saturation, because the rectangular hyperbola does not allow a rapid enough transition from light-limited to light-saturated photosynthesis. A modified rectangular hyperbola (Eq. 7.3, Table 7.1) provides a much better fit to the experimental data by using higher order terms in the denominator, but this equation has no greater mechanistic justification than any of a number of other possible formulations.

The most popular models currently employed to describe experimental PI curves are the hyperbolic tangent function (Eq. 7.8, Table 7.1) and an exponential model (Eq. 7.6). The hyperbolic tangent function was originally introduced as a purely empirical function, but has found theoretical justification in an analysis by Chalker (1980). The exponential function is mathematically identical to the cumulative one-hit Poisson distribution used to describe the relationship between flash yield and flash intensity (Mauzerall and Greenbaum, 1989).

Jassby and Platt (1976) reformulated eight models of gross photosynthesis, so that they were expressed in terms of the same two parameters (Table 7.1). These parameters are α, defined here as the light-limited initial slope of the PI curve at low incident irradiances and P_m, the light-saturated maximum photosynthesis rate. To account for negative values of net photosynthesis in darkness, the authors needed a third parameter, R, the dark respiration rate. The various formulations of the PI curve differ in the abruptness of the transition from light-limited to light-saturated photosynthesis. When P_m, α, and R are assumed to have fixed values, the Blackman formulation (Eq. 7.1) greatly overestimates photosynthesis and the rectangular hyperbola (Eq. 7.2) greatly underestimates photosynthesis at intermediate light levels, relative to the other six models.

The accuracy with which the different models describe the PI curve has been considered for free-living and symbiotic microalgae and macroalgae. Working with ^{14}C-uptake data for natural assemblages of marine phytoplankton, Jassby and Platt (1976) concluded that the hyperbolic tangent function (Eq. 7.8) provided the best fit to the data. However, a reanalysis of their data with improved curve-fitting techniques indicated that most models (i.e., Eqs. 7.3, 7.5, 7.6, 7.7, and 7.8) could not be distinguished on a criterion of goodness of fit, although Blackman kinetics and the Michaelis-Menton equation were clearly inferior descriptions of the PI curve (Lederman and Tett, 1981). Chalker (1981) concluded that the hyperbolic tangent function (Eq. 7.8) and the exponential function (Eq. 7.6) provided more accurate estimates of α and P_m in the reef-building corals than did Blackman kinetics or the rectangular hyperbola, consistent with Lederman and Tett's (1981) analysis for phytoplankton. Coutinho and Zingmark (1987) reported that the hyperbolic tangent provided the best fit to PI curves of marine macroalgae based on both the criteria of goodness of fit and minimizing residuals. Surprisingly, a recent investigation of the PI curve in marine macroalgae found poor fit and low precision in estimates of α and P_m using the hyperbolic tangent function (Nelson and Siegrist, 1987). Researchers involved in terrestrial plant photosynthesis have circumvented the problem of choosing the most appropriate model by introducing the concept of convexity, in which the shape of the PI curve is determined by a parameter, which, like P_m or α, needs to be independently assessed (Leverenz, 1987). In the first application of the concept of convexity to algae, Leverenz et al. (1990) demonstrated that the shape of the PI curve depended on the degree of photoinhibition in *Chlamydomonas reinhardtii*. The transition from light-limitation to light-saturation became less abrupt with increased photoinhibition.

Alternative Forms of the Photosynthesis–Light Response Curve

Most investigators express photosynthesis rates on a unit chlorophyll *a* basis because chlorophyll *a* is a readily measured index of the size of the photosynthetic apparatus. However, some investigators prefer to express photosynthesis rates in terms of the optical cross section and turnover time of a photosynthetic unit (PSU) rather than in terms of the chlorophyll *a* specific maximum rate P_m^{chl} and the initial slope α^{chl}. The analysis of steady-state photosynthesis in terms of PSUs is derived from the treatment of O_2 flash yield versus flash irradiance curves (Ley and Mauzerall, 1982; Mauzerall and Greenbaum, 1989).

$$Y/Y_m = 1 - \exp(-\sigma^{PSU}E) \tag{7.9}$$

where Y is the yield of O_2 per flash, Y_m is the maximum yield of O_2 in a saturating flash, σ^{PSU} is the 'optical' cross-section of a PSU and E is the flash irradiance (μmol photons·m^{-2}·flash^{-1}). In this section, we show that the two approaches are equivalent when an exponential model (i.e., Eq. 7.4) is employed. Following Mauzerall and Greenbaum (1989), the optical cross-section (σ^{PSU}) can be considered to equal the product of the maximum photon yield for O_2 evolution, the *in vivo* absorption coefficient of a chlorophyll *a* molecule and the size of the PSU:

$$\sigma^{PSU} = \phi_m a^*/N^{PSU} \tag{7.10}$$

where ϕ_m is the maximum photon yield (mol $O_2 \cdot mol^{-1} \cdot$photons), a^* is the *in vivo* absorption coefficient normalized to chlorophyll a and N^{PSU} is the amount of chlorophyll a in a PSU. Thus, σ^{PSU} is analogous to the initial slope of the PI curve when photosynthesis is normalized to PSUs rather than to chlorophyll a; specifically,

$$\sigma^{PSU} = \phi_m a_{PSU}^* = \Phi_m a^* N^{PSU} \qquad (7.11)$$

where a_{PSU} is the specific absorption coefficient normalized to a photosynthetic unit.

Based on Mauzerall's treatment of the flash yield versus flash irradiance curve, Dubinsky et al. (1986) proposed the following form for a continuous light PI curve:

$$P/P_m = 1 - \exp(-\sigma^{PSU} \tau E) \qquad (7.12)$$

where σ^{PSU} is the optical cross section of a PSU, τ is the turnover time of a PSU in continuous light, and E is the irradiance ($\mu mol \cdot m^{-2} \cdot s^{-1}$). Noting that by definition $P_m^{PSU} = 1/\tau$ and $\sigma^{PSU} = \alpha^{PSU}$ allows Eq. 7.18 to be rewritten as

$$P^{PSU} = P_m^{PSU} (1 - \exp(\alpha^{PSU} E/P_m^{PSU}) \qquad (7.13)$$

This is mathematically analogous to Eq. 7.14. By noting the identities

$$\begin{aligned} P^{chl} &\equiv P^{PSU} N^{PSU} \\ \alpha^{chl} &\equiv \alpha^{PSU} N^{PSU} \\ P_m^{chl} &\equiv P_m^{PSU} N^{PSU} \end{aligned} \qquad (7.14)$$

Equation 7.6 is readily obtained from Eq. 7.13.

One advantage gained by expressing both the initial slope and light-saturated rate of photosynthesis in terms of the number of photosynthetic units is the explicit recognition that α and P_m are not necessarily functionally independent. The often-observed covariation of α and P_m (Harding et al., 1987) can be most readily explained in terms of changes in the number of active photosynthetic units, all other factors remaining unchanged. When α and P_m vary independently, it becomes necessary to invoke changes in τ, a_{PSU}^*, or ϕ_m.

Mechanistic Interpretation of the
Photosynthesis-Light Response Curve

As described above, the parameters of the PI curve are strictly empirical coefficients. To gain a greater understanding of the environmental or physiologic factors that regulate photosynthesis, it is desirable to relate these parameters to more fundamental processes. For the remainder of this discussion, we assume that the PI curve can be described by the exponential function:

$$P^{chl} = P_m^{chl} [1 - \exp(-\alpha^{chl} E/P_m^{chi})] \tag{7.15}$$

Both the initial slope (α^{chl}) and the light-saturated rate (P_m^{chl}) have been described in terms of underlying biochemical and biophysical events.

At low irradiance, the photosynthesis rate is proportional to the amount of light intercepted by the photosynthetic apparatus, and the proportionality constant is the initial slope α^{chl}. The initial slope can thus be equated to the product of a chlorophyll a-specific absorption coefficient and the maximum realized photon yield of photosynthesis (ϕ_m):

$$\alpha^{chl} = a^*\phi_m \tag{7.16}$$

Absorption coefficients are often defined per unit chlorophyll a but can also be normalized to units of per cell or per unit mass.

Much of the variation in α^{chl} in actively growing microalgae appears to be due to variations in light absorption (a^*), rather than to variations of the photon yield (ϕ_m) (Welschmeyer and Lorenzen, 1981). However, the initial slope has been observed to decrease under adverse growth conditions due to a decline of ϕ_m (Welschmeyer and Lorenzen, 1981; Cleveland and Perry, 1987), and interspecific variations of ϕ_m have been reported (Dubinsky et al., 1986). During phenotypic responses to irradiance or nutrient limitation, compensating changes in ϕ_m and a^* may lead to the situation where there is less variation in α^{chl} than in either ϕ_m or a^* (Dubinsky et al., 1986; Herzig and Falkowski, 1989).

Typically, the photon yield is calculated from measurements of the initial slope and the absorption coefficient by rearranging Eq. 7.16 (Geider et al., 1985; Dubinsky et al., 1986). Only in the experiments of Welschmeyer and Lorenzen (1981) and Cleveland and Perry (1987) has ϕ_m been obtained directly from simultaneous measurements of absorption and photosynthesis using an integrating sphere (i.e., from the initial slope of a plot of photosynthesis versus absorbed light). Whether measured directly or indirectly, all estimates of ϕ_m are subject to potential errors in the measurements of light absorption outlined in Chapter 6. In addition, variations in gas exchange rates due to induction phenomena (Walker, 1981), changes in mitochondrial respiration between light and darkness (Kok, 1949), and changes in gas exchange with irradiance due to photorespiration produce ambiguities that confound an accurate estimate of ϕ_m (Pirt, 1986; Osborne and Geider, 1987a). With material acclimated to low irradiances or nutrient-deficient conditions, light-limited rates of photosynthesis may only be marginally greater than the light compensation point, making it difficult to estimate the initial slope accurately.

These uncertainties are manifested by a continuing controversy over the absolute upper limit of ϕ_m (Pirt, 1986; Osborne and Gelder, 1987a). According to the Z-scheme for photosynthesis, a minimum of eight photons are required per molecule of O_2 evolved, giving a maximum upper limit for ϕ_m of 0.125 mol $O_2 \cdot mol^{-1} \cdot$ photon. The value of 0.125 is an upper limit, which is unlikely to be realized in practice because of expected inefficiencies in photosynthetic electron transport (Myers, 1980; Bell, 1985). Although many investigators report values for $\phi_m < 0.125$, a number of investigations are inconsistent with the Z scheme in that values of $\phi_m > 0.125$ have been observed (Brackett et al., 1953b; Pirt, 1986; Osborne and Geider, 1987a). In view of recently expressed reservations about the operation of the two photosystems (Duysens, 1989) and the demonstration of the direct photoreduction of $NADP^+$ by PS2 (Arnon and Tang, 1989; Arnon and Barber, 1990), additional experimental determinations of ϕ_m under well-controlled conditions that are free of the experimental artifacts discussed above are clearly warranted.

Largely as a consequence of problems associated with transitory gas exchange phenomena, some workers have suggested that the maximum value for ϕ_m can only be estimated with actively growing cells in balanced growth (Pirt, 1986; Osborne and Geider, 1987a). Measurements of ϕ_m based on balanced growth are technically more difficult than estimates based on gas exchange, because of the need to maintain stable conditions of biomass and irradiance over several division cycles (i.e., days to weeks). It is, therefore, not surprising that only a few studies (Göbel, 1987a, b; Myers, 1980; Pirt, 1986; Osborne and Geider, 1987b) have attempted to estimate ϕ_m from measurements of light absorption and growth. Unfortunately, even these measurements do not provide unequivocal estimates of ϕ_m because of difficulties associated with the determination of metabolic costs associated with cell synthetic activity.

A finite amount of time is required for transfer of an electron from H_2O to CO_2, and that time interval (τ) is referred to as the turnover time of a PSU. Light saturation of photosynthesis occurs when τ exceeds the time interval between successive excitations. Photosynthetic unit size was originally defined in terms of the number of PS2 reaction centers, determined from oxygen flash yield experiments (Emerson and Arnold, 1932),

Table 7.2. Stoichiometry of electron transport chain components and RUBISCO in microalgae.

Species	Irradiance $\mu mol \cdot m^{-2} \cdot s^{-1}$	RCII:cytb/f:RCI			RUBISCO
Chlamydomonas	47	1.3	1.4	1.0	—
reinhardtii	200	2.2	—	1.0	—
	400	2.6	2.8	1.0	—
Chlorella	low	1.4	—	1.0	—
pyrenoidosa	high	1.3	—	1.0	—
Dunaliella tertiolecta	45	0.9	—	1.0	—
	600	0.8	—	1.0	—
Dunaliella tertiolecta	80–1,900	0.8	1.4	1.0	0.9–3.5
Mantoniella	5	3.7	1.2	1.0	—
squamata	100	2.4	1.8	1.0	—
Isochrysis galbana	30–600	2.0	—	1.0	—
Isochrysis galbana	(N-limitation)	1.2	4.0	1.0	5.3–14.4
Skeletonema	30	2.3	—	1.0	—
costatum	200	1.4	—	1.0	—
	600	1.1	—	1.0	—
Ny17	27	0.89	—	1.0	—
	274	0.76	—	1.0	—
UP45	27	0.44	—	1.0	—
	274	0.52	—	1.0	—
Skel	27	0.85	—	1.0	—
	274	0.63	—	1.0	—
Thalassiosira weisflogii	30–600	2.0	—	1.0	—
Prorocentrum micans	70–600	1.4	—	1.0	—

Data were obtained from the following sources: Dubinsky et al. (1986), Neale and Melis (1986), Falkowski et al. (1981, 1989), Myers and Graham (1983), Sukenik et al. (1987), Gallagher et al. (1984), and Wilhelm et al. (1989).

and, until recently, it was assumed that the ratio of PS2:cyt b_6/f:PS1 was fixed at unity, based on the Z scheme for photosynthetic electron flow. It is now known that the photosynthetic electron transport chain is less rigidly defined than is required by a literal interpretation of the Z scheme, and various stoichiometries of reaction centers and electron transport chain catalysts have been observed (Table 7.2). When dealing with intact cells, the photosynthetic unit can be empirically defined as all of the catalysts or compounds associated with a PS2 reaction center that are required for whole-chain electron flow from H_2O to CO_2, resulting in the evolution of one O_2 molecule. Thus, P_m^{chl} can be expressed as:

$$P_m^{chl} = 1/\tau N^{PSU} \tag{7.17}$$

One disadvantage of describing whole-chain electron flow in terms of photosynthetic unit number and turnover time of an entire PSU is that this approach avoids the question of what process actually sets the limit on PSU turnover during steady photosynthesis in continuous light. Light-saturated photosynthesis, as determined by net gas exchange techniques, may be set by reactions downstream and to some extent independent of photosynthetic electron transport (Stitt, 1986; Sharkey et al., 1986a, b; Sukenik et al., 1987). Extending the concept of the photosynthetic unit to include the catalysts of these downstream reactions would greatly dilute the original intent in defining the PSU in terms of a component of photosynthetic light harvesting and electron transport. Use of ^{18}O techniques to obtain true rates of gross oxygen evolution should lead to an increased understanding of the factors that regulate light-saturated photosynthesis. Significant rates of light-dependent oxygen consumption at light saturation may indicate an excess of electron transport capacity over carboxylation capacity.

Photosynthesis at Irradiances near the Light Compensation Point

The light compensation point is that irradiance at which net gas exchange is zero because gross photosynthesis and respiration rates are in balance. The compensation irradiance can be related to the initial slope and respiration rate as follows:

$$E_0 = R^{chl}/\alpha^{chl} \tag{7.18}$$

where E_0 is the compensation irradiance (μmol photons·m^{-2}·s^{-1}); R^{chl} is the respiration rate (μmol O_2·mg chla^{-1}·s^{-1}) and, α^{chl} is the initial slope of the PI curve based on chlorophyll content (mol O_2·mol^{-1} photons·m^{-2}·mg^{-1} chl$a$). The compensation irradiance is not fixed but varies primarily due to R^{chl} and to a lesser extent α^{chl}, both of which will depend on the environmental conditions (irradiance, nutrient availability, temperature), under which the algae were growing.

A careful examination of photosynthesis rates at irradiances near the compensation point often reveals a break in the initial slope, in which the slope abruptly increases at low irradiances (Fig. 7.2). The change in slope is referred to as the Kok-effect in recognition of the contributions of the investigator who first studied this phenomenon (Kok, 1948, 1949, 1951). The Kok effect can be overlooked if the PI curve is poorly resolved near the compensation irradiance and is ignored in all mathematical treatments of the PI curve, such as those summarized in Table 7.1.

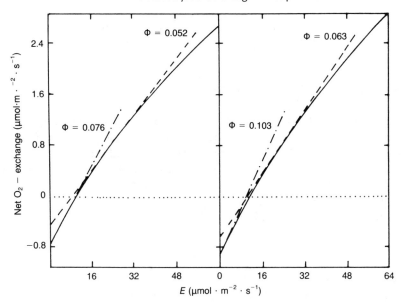

Fig. 7.2. The response of photosynthesis to limiting irradiance in the cyanobacteriium *Anacystis nidulans* previously grown at 200 $\mu mol \cdot m^{-2} \cdot s^{-1}$ to indicate the influence of the Kok effect on the photon yield ($\emptyset_m$) of photosynthesis in the absence (a) or presence (b) of nitrate. Figures next to the dashed lines in (a) and (b) represent the calculated slopes of the linear least-squares regression analysis of the photon yield above (---) and ($\cdots$) below the light compensation point. Redrawn from Romero et al. (1989).

However, the Kok effect may provide considerable insight into the relationship between photosynthetic and respiratory metabolism. Kok (1948) found that the ratio of the slopes below and above the break in *Chlorella* varied with physiological condition by up to a factor of two; in the extreme case, an extrapolation of the upper line to zero irradiance gave an estimate of respiration that was only half that of the observed dark respiration rate. Kok (1948) initially interpreted the change in slope as a reflection of changes in the photon yield of photosynthesis (ϕ_m). However, it is now thought that the higher slope below the compensation point reflects a partial light inhibition of the respiration rate at low irradiances (Sharp et al., 1984). The Kok effect is not always observed (Emerson and Chalmers, 1957), suggesting that light-inhibition of respiration may be transitory or only occur under certain physiological conditions. It has also been suggested that the efficiency of metabolite transport across the chloroplast envelope and the rate of cytosolic ATP consumption may determine the magnitude of the Kok effect (Heber et al., 1982). Osborne and Geider (1987a) hypothesized that the Kok effect would be particularly evident following addition of nitrate to nitrogen-starved cells, as has recently been confirmed for the cyanobacterium *Anacystis nidulans* (Romero et al., 1989 Fig. 7.2).

Models That Include Photoinhibition

Reductions of biomass-specific photosynthesis rates in samples incubated at or near the surface are often observed under in situ or simulated in situ conditions (Harris, 1978; Ramus and Rosenberg, 1980). Equations that account for this high irradiance inhibition of photosynthesis have been developed by several investigators working in both fresh and salt waters. Platt et al. (1980) extended the exponential formulation (Eq. 7.6) to account for photoinhibition as follows:

$$P = P_s [1 - \exp(-aE)] \exp(-bE) \qquad (7.19)$$

where $a = \alpha/P_s$ and $b = \beta/P_s$. In this formulation P_s is not necessarily equivalent to the P_m of Eq. 7.6. When photoinhibition is not observed $P_m = P_s$. However, when photoinhibition is observed $P_s > P_m$. Furthermore, the value of α determined from just the nonphotoinhibited portion of the PI curve will equal the sum $\alpha + \beta$ determined from a complete PI curve that includes photoinhibition. The three-parameter model of gross photosynthesis works well in many environments, but fails to adequately describe the extended plateau observed in some samples. Coutinho and Zingmark (1987) report that Eq. 7.19 provides a good fit to those PI curves of marine macrophytes that exhibit photoinhibition.

Fasham and Platt (1983) derived a four-parameter model of gross photosynthesis, including photoinhibition, which allowed more flexibility than previous models. Although derived from a description of the light reactions of photosynthesis that are typified by time constants of less than a second, the model was applied to data obtained from 4 to 6-hr incubations. This mismatch in time scales between the mechanistic processes incorporated into the model and the observations to which the model was applied may limit the usefulness of this model in other situations.

Megard et al. (1984) employed a three-parameter model that extended the rectangular hyperbola to include photoinhibition through addition of an E^2 term in the denominator,

$$P = P_s / (K_1 + E + E^2/K^2) \qquad (7.20)$$

This equation is identical to Haldane's (1930) formulation for enzyme reactions that are inhibited by high substrate concentrations. A mechanistic derivation involving light-dependent activation and inactivation of PS2 was provided by Megard et al. (1984), and a somewhat different formulation of the same model was proposed by Peeters and Eilers (1978).

Reduced gas exchange rates at high irradiance are due largely to two processes: photoinhibition and photorespiration. Photoinhibition is a time-dependent process that appears to involve a reversible or irreversible inactivation of PS2 (Neale, 1987). Photorespiration is caused by the oxygenase reaction of ribulose bisphosphate carboxylase-oxygenase, which is increased under high light and/or reduced CO_2 supply or increased O_2 concentration (Badger, 1985). An additional oxygen-consuming reaction in illuminated plants, the Mehler reaction, may play a role in the protection of the photosynthetic apparatus from photoinhibition (Badger, 1985).

High light inhibition of photosynthesis occurs on a time scale that is commensurate with the time required to determine the PI curve. It is thus likely that models that characterize high light inhibition of photosynthesis by a single, time-invariant parameter may not adequately describe the light-dependence of photosynthesis on shorter time scales. This question has been addressed by Gallegos and Platt (1985) and Neale (1987).

Interspecific and Intraspecific Variations in P_m and α

Photosynthesis rates are normalized to some measure of biomass to allow comparisons between organisms with widely different sizes and thus to account for the general observation that the rate of photosynthesis scales with plant biomass. The choice of a biomass unit for normalizing photosynthesis rates is often based on convenience, but this choice is important in making interspecific and intraspecific comparisons (Myers, 1970; Talling, 1984). In microalgae, photosynthesis rates are often expressed as cell-specific or chlorophyll-specific rates or more rarely as carbon-specific or mass-specific rates. In macroalgae, photosynthesis is often normalized to unit area, dry weight or chlorophyll, and occasionally to unit wet weight or carbon content. The PI curve will have the same shape whatever the unit of biomass to which photosynthesis rates are normalized. However, changes in the absolute values of biomass-specific photosynthesis rates in response to environmental factors will depend on the unit of biomass to which photosynthesis rates have been normalized.

One goal of algal photosynthesis research is to determine the magnitude of phenotypic and genotypic variations in the parameters of the photosynthesis–light response curve. Phenotypic variation can be examined by comparing the attributes of genetically identical organisms grown under a range of environmental conditions, whereas genotypic variation is investigated by comparing photosynthesis and biomass of a range of genetically unique organisms grown under identical and well-defined conditions. Phenotypic responses to irradiance have been best characterized, whereas responses to light quality, nutrient availability, and temperature are not well understood. The effects of both phenotypic and genotypic variability of the photosynthetic apparatus of algae have rarely been examined in the same investigation, although the work of Gallagher et al. (1984) is a notable exception. The remainder of this section attempts to summarize our current understanding of the influence of phenotypic and genotypic variations on photosynthetic physiology.

Photoadaptation of photosynthesis

An examination of the response of the photosynthetic apparatus to irradiance, nutrient availability, or temperature requires that all other factors be controlled. Photosynthetic parameters can vary widely in batch cultures of marine microalgae (Griffiths, 1973; Beardall and Morris, 1976), demonstrating the importance of ensuring that cells are in balanced growth (Shuter, 1979) before attempting to examine photoadaptation. However, it may take 5–20 generations before phytoplankton become fully acclimated to the prevailing environmental conditions (Brand et al., 1981). It is arguable that balanced growth is never achieved in batch cultures, because the organisms are continuously modifying their environment, which will in turn modify growth processes (Wanner and Egli, 1990). Turbidostats can be used to maintain cells in light-limited balanced growth (Falkowski et al., 1985) and chemostats can be used to maintain cells in nutrient-limited balanced growth (Droop, 1966). However, it is possible to maintain cells in nearly balanced growth in both nutrient-sufficient and nutrient-limited cultures by frequent subculturing (Brand et al., 1981).

Algae commonly respond to reduced irradiance by increasing the quantity of photosynthetic pigments. This can be viewed as an adaptive response in which the components of the light harvesting apparatus increase in abundance when light is a limiting resource. In nutrient-sufficient microalgae, a plot of the carbon-to-chlorophyll a ratio against irradiance is often linear, although the relationship is sensitive to temperature, with higher carbon-to-chlorophyll a ratios at lower temperatures (Geider, 1987). In addition to changes in the size of the photosynthetic apparatus, as indicated by lower carbon-to-chlorophyll a ratios, the response to irradiance may involve changes in the composition of the photosynthetic apparatus. The most common response to low irradiance is an increase in the antennae size of the photosynthetic unit, which is reflected in an increase in the ratio of total chlorophyll a to reaction center P_{700} (Perry et al., 1981).

The classic pattern for the acclimation of the PI curve to irradiance is Myer's (1970) data for *Chlorella pyrenoidosa* (Fig. 7.3). To understand this response, it is necessary to consider both chlorophyll a-specific and dry-weight-specific photosynthesis rates. Myers (1970) found that the chlorophyll a-specific initial slope of the PI curve was largely independent of the irradiance at which the cells were grown but the chlorophyll a-specific light-saturated rate (P_m^{chl}) decreased at low irradiance. In contrast, the dry-weight-specific light-saturated photosynthesis rate (P_m^w) was less dependent on irradiance, but the initial slope (α^w) decreased at low irradiance.

Myer's (1970) observations (Fig. 7.3) suggest that light-limited photosynthesis is controlled by the efficiency of light absorption, which covaries with pigment concentration. What controls light-saturated photosynthesis is less clear, but it appears to be the concentration of some component(s) which covary(ies) with algal mass. Foy and Gibson (1982b) found that cell-protein-specific P_m was independent of irradiance in *Oscillatoria agardhii*, whereas dry-weight-specific P_m decreased at high irradiance, presumably because of the accumulation of noncatalytic carbohydrate storage reserves. It is not surprising that cell protein provides a good predictor of light-saturated photosynthesis given the large contribution of RUBISCO and thylakoid proteins to total cell protein (Falowski et al., 1989).

Other patterns in the response of chlorophyll a-specific and weight-specific photosynthesis rates have also been described. Richardson et al. (1983), for example, list five different types of response of the PI curve to preconditioning irradiance. It is possible, however, that the range of responses reported by Richardson et al. (1983) is not an inherent property of different algal groups but represents different components of a continuum of photoadaptation along a gradient from light-limitation to photoinhibition.

Chlorophyll-specific, light-saturated photosynthesis rates generally increase in microalgae acclimated to high irradiance. Two- to fivefold changes of P_m^{chl} in response to differences in irradiance have been documented in a wide range of microalgae, including chlorophytes (Myers and Graham, 1971; Falkowski and Owens, 1980), diatoms (Beardall and Morris, 1976; Falkowski and Owens, 1980; Geider et al., 1984, 1985), dinoflagellates (Prézelin, 1976; Prézelin and Sweeny, 1978), cyanobacteria (Foy and Gibson, 1982a, b; Raps et al., 1983), and a rhodophyte (Levy and Gantt, 1988). An almost linear increase of P_m^{chl} with the logarithm of

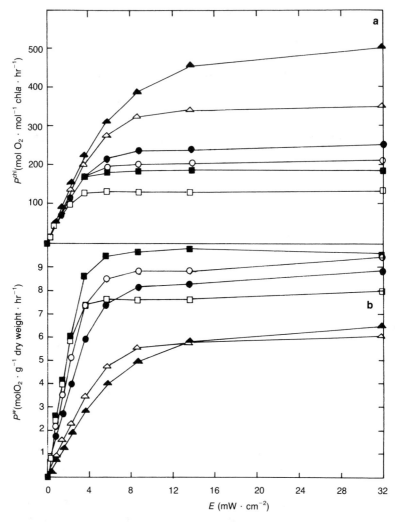

Fig. 7.3. Response of the photosynthesis–irradiance (PI) curve to light-limited growth in *Chlorella pyrenoidosa*. Data of Myers and Graham (1971). Individual PI curves for cells cultured at nutrient-sufficient, but light-limited growth rates of 0.35 d^{-1} (□), 0.78 d^{-1} (■), 1.3 d^{-1} (○), 1.8 d^{-1} (●), 2.3 d^{-1} (△). Chlorophyll *a*-specific rates (a) show little variation in the initial slope but approximately a fivefold variation in the light-saturated rate (P_m^{chl}). In contrast, dry weight-specific rates (b) show much larger variations in the initial slope but only approximately a twofold variation in the light-saturated rate (P_m^w).

172

irradiance is observed in several diatoms grown under nutrient-sufficient conditions in continuous light (Fig. 7.4a).

In contrast to P_m^{chl}, α^{chl} generally varies little with preconditioning growth irradiance (Fig. 7.4b). For example, Foy and Gibson (1982a) found small changes in α^{chl} in response to an increase in growth irradiance from 30 to 150 $\mu mol \cdot m^{-2} \cdot s^{-1}$ in 20 strains of freshwater cyanobacteria. The value of α^{chl} increased up to 20% in 5 strains, decreased up to 25% in 15 strains, with a mean overall decrease of 6% when data from all 20 strains are considered. The phenotypic variability of α^{chl} in response to preconditioning irradiance is generally limited to less than a twofold range, and often differences of less than 25% are observed (Myers and Graham, 1971; Prézelin, 1976; Prézelin and Sweeney, 1978; Foy and Gibson, 1982; Raps et al., 1983; Geider et al., 1984, 1985; Dubinsky et al., 1986; Levy and Gantt, 1988).

Carbon-specific, light-saturated photosynthesis rates (P_m^c) can be considered as the product of P_m^{chl} and chl$a$:C. Because P_m^{chl} and chl$a$:C have the opposite dependencies on irradiance, it is not immediately evident how P_m^c will vary with growth irradiance. Some species show a small dependence of P_m^c on irradiance. In *Chlorella pyrenoidosa*, P_m^c varied approximately 1.5-fold, despite a 20-fold change in irradiance and a 7-fold change in μ, whereas P_m^c did not change in *Glenodinium* sp., despite a 10-fold change in irradiance and a 3-fold change in μ (Prézelin, 1976). However, other species show a pronounced decline in P_m^c at low μ (Falkowski and Owens, 1980; Geider et al., 1984, 1985), and/or a decrease as μ approaches μ_m (Dubinsky et al., 1986). An interspecific comparison suggests that P_m^c is correlated with the maximum resource-unlimited growth rate (Fig. 7.5).

Diel Patterns of Photosynthesis

Pronounced diel oscillations in phytoplankton photosynthesis (Doty and Oguri, 1957; Sournia, 1974) complicate the interpretation of the response of the photosynthetic apparatus to variations in irradiance. Diel periodicities are evident in both light-saturated and light-limited photosynthesis rates, with P_m and α undergoing proportional changes in magnitude in both laboratory cultures of marine microalgae (Prézelin and Sweeney, 1977; Harding et al., 1981, 1987), and in phytoplankton populations collected from the field (MacCaull and Platt, 1977; Harding et al., 1982). This covariation of P_m and α is one of the major patterns that emerges in photosynthesis experiments and may indicate the activation/inactivation of photosynthetic units by endogenous or environmental factors in situ. Diel patterns of P_m and α commonly show maxima during the light period and minima in darkness, although the timing of maximum α and P_m

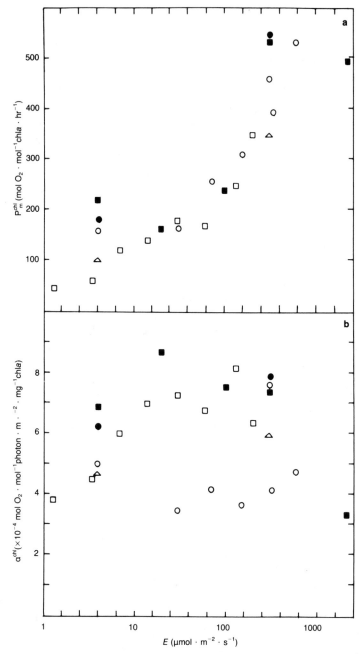

Fig. 7.4. Responses of (a) chlorophyll-specific, light-saturated photosynthesis (P_m^{chl}) and (b) the initial slope (α^{chl}) of the PI curve to preconditioning irradiance. Note the logarithmic transformation of irradiance on the abscissa. Data for *Phaeodactylum tricornutum* (□) from Geider et al. (1985, 1986), *Thalassiosira weisflogii* (○) from Perry et al. (1981) and Dubinsky et al. (1986), *Thalassiosira pseudonana* (■) from Perry et al. (1981) and Cullen and Lewis (1988), *Chaetoceros danicus* (△) from Perry et al. (1981), and *Chaetoceros gracilis* (●) from Perry et al. (1981). A photosynthetic quotient of unity was assumed in converting the ^{14}C data of Perry et al. (1981) and Cullen and Lewis (1988) to the equivalent O_2 evolution rates.

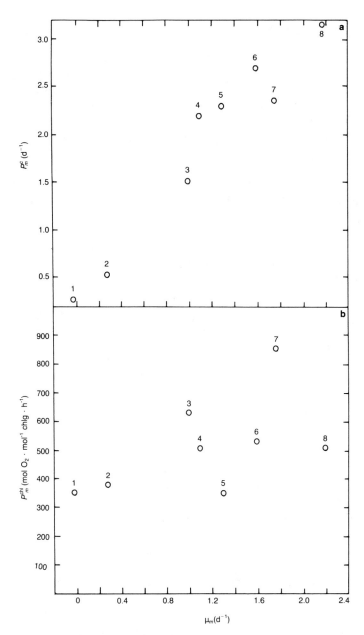

Fig. 7.5. Response of carbon-specific, light-saturated photosynthesis to growth rate (μ_m), in light-and nutrient-saturated cultures. P_m^c is linearly related to the resource-unlimited growth rate suggesting that the potential for growth may limit the maximum biomass-specific rate of photosynthesis. In (b) the response of chlorophyll a-specific, light-saturated photosynthesis to μ_m shows that P_m^{chl} is largely independent of μ_m. Data are shown for (1) *Prorocentrum micans*, (2) *Prochlorothrix hollandica*, (3) *Isochrysis galbana*, (4) *Microcystis aeruginosa*, (5) *Phaeodactylum tricornutum*, (6) *Thalassiosira weisflogii*, (7) *Synechococcus* clone DC2, and (8) *Chlorella pyrenoidosa*, from the investigations of Myers and Graham (1971), Raps et al. (1983), Geider et al. (1985), Falkowski et al. (1985), Dubinsky et al. (1986), Kana and Glibert (1987), Burger-Wiersma and Post (1989).

varies between midmorning and midafternoon (MacCaull and Platt, 1977; Forbes et al., 1986; Harding et al., 1987). In contrast to the observations for microalgae, in which both P_m and α exhibit diel periodicities, Mishkind et al. (1979) found that only P_m exhibited a diel variation in *Ulva lactuca*, whereas α did not. The diel variation in P_m was attributed to an endogenous rhythm in the rate of reoxidation of plastoquinone by PS1 (Mishkind et al., 1979).

In a comprehensive examination of 24 marine phytoplankton species grown at 15°C and 1,200 $\mu W \cdot m^{-2}$ (approximately 60 $\mu mol \cdot m^{-2} \cdot s^{-1}$) on a 12:12 light-dark cycle, Harding et al. (1981) documented diel variations in 17 species. A 150-fold difference was observed for P_m in the diatom *Ditylum brightwellii*, and a 50–60-fold difference was observed in *Lauderia borealis*. Of the species that exhibited no diel periodicity, most were small and fast growing and these included three common laboratory "weeds" (i.e., the diatoms *Phaeodactylum tricornutum* and *Skeletonema costatum* and the chlorophyte *Dunaliella tertiolecta*). In a subsequent investigation using *D. brightwellii*, Harding et al. (1981) showed that diel periodicity in the PI curve was most pronounced in early exponential phase cultures where cell-specific maximum photosynthesis rates (P_m^{cell}) varied from < 10 to 170 pg $C \cdot cell^{-1} \cdot h^{-1}$, and was least evident in late stationary phase cultures, in which P_m^{cell} varied from 4.5 to 7.5 pg $C \cdot cell^{-1} \cdot h^{-1}$.

Nutrient dependence of photosynthesis

The components of the photosynthetic apparatus, namely the pigment-protein complexes, the carriers of the photosynthetic electron transport chain, and enzymes of the photosynthetic carbon reduction cycle, account for a large fraction of total cell nitrogen in micro- and macroalgae. For example, Falkowski et al. (1989) calculated that 20 to 30% of total cell nitrogen could be attributed to the carboxylating enzyme RUBISCO in nutrient-sufficient *Isochrysis galbana*. The nitrogen costs of the light-harvesting complexes were determined by Raven (1984) to range from 28 mol nitrogen·mol^{-1} chromophore in the chlorophyll a and b complexes of the chlorophyta to 197 mol nitrogen·mol^{-1} chromophore associated with allophycocyanin in the phycobiliphytes. It is therefore not surprising that algae respond to a reduction in nitrogen availability by reducing the size of the photosynthetic apparatus. This reduction in pigmentation may be adaptive, if it allows a redistribution of nitrogen from light harvesting proteins to other catalysts or if it affords greater protection from photoinhibition when growth processes limit the rate of consumption of the products of photosynthesis. However, a reduction in the chlorophyll a: carbon ratio (chlorosis) appears to be a general response to nutrient limitation, and is not specific to nitrogen limitation (Goldman,

1980; Laws and Bannister, 1980; Osborne and Geider, 1986; Herzig and Falkowski, 1989) and macroalgae (Lapointe and Duke, 1984). The range of chl$a$:C encountered is likely to depend on both irradiance and temperature, with low temperature and high irradiance contributing to decreases of chl$a$:C, which are most pronounced at low nutrient-limited growth rates.

In addition to a role as light-harvesting accessory pigments, phycobiliproteins have been suggested to be a major intracellular nitrogen reserve in cyanobacteria (Boussiba and Richmond, 1980) and red algae (Lapointe and Duke, 1984). Although efficiently coupled to photosynthesis in low light environments, phycobiliproteins may become uncoupled from photosynthesis under high irradiances. Phycobiliproteins decrease to a greater extent than chlorophyll a under nitrogen-limited conditions with much of the nitrogen in the phycobiliproteins presumably being mobilized to allow continued growth in a nitrogen-limited environment (De Loura et al., 1987). In contrast to the phycobiliphytes in which the accessory pigment-to-chlorophyll a ratio decreases under nutrient-limited conditions, both chlorophyll c:chlorophyll a and carotenoid:chlorophyll a increased with increasing N-limitation in *Isochrysis galbana* (Herzig and Falkowski, 1989).

Chlorophyll a-specific, light-saturated photosynthesis rates typically decline slightly under moderate nutrient-limitation, dropping off markedly only as μ approaches zero (Osborne and Geider, 1986; Herzig and Falkowski, 1989). A linear dependence of carbon-specific, light-saturated photosynthesis rates on nutrient-limited growth rate has been observed in several microalgae (Fig. 7.6) and a macroalga (Lapointe and Duke, 1984). Correlated with the reduction in P_m^c is a decrease in the proportion of cell nitrogen, which is associated with RUBISCO (Falkowski et al., 1989). This response to nutrient limitation involves large reductions in chlorophyll per cell and slight reductions in the chlorophyll-specific light-saturated photosynthesis rate (Osborne and Geider, 1986; Lapointe and Duke, 1984; Falkowski et al., 1989).

In general, chlorophyll a-specific rates of light-limited photosynthesis are less affected by nutrient limitation than are light-saturated rates. The initial slope was independent of growth rate in phosphorus-limited chemostat cultures of *Scenedesmus quadricauda* (Smith, 1983) and phosphorus starved batch cultures of *Anabaena wisconsinense* and *Chlorella pyrenoidosa* (Senft, 1978). However, α^{chl} increased slightly with increasing nitrate-limitation in *Phaeodactylum tricornutum* (Osborne and Geider, 1986), and *Isochrysis galbana* (Herzig and Falkowski, 1989). A large increase in the chlorophyll a-specific absorption coefficient was partially offset by decreases in the maximum photon yield of photosynthesis in these microalgae. Osborne and Geider (1986) observed that α_{chl} increased

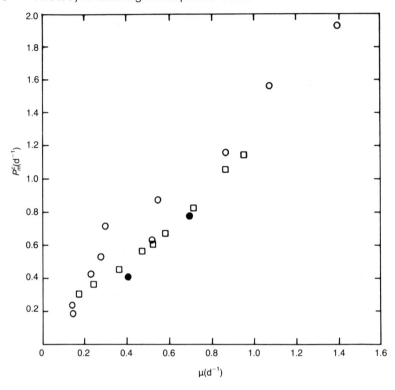

Fig. 7.6. Variation in carbon-specific, light-saturated photosynthesis (P_m^c, with units of d^{-1}) as a function of growth rate (μ, in units of d^{-1}) for nutrient-limited phytoplankton. Data for *Thalassiosira weisflogii* (●) from Terry (1982), *Phaeodactylum tricornutum* (○) from Terry (1982) and Osborne and Geider (1986), and *Isochrysis galbana* (□) from Herzig and Falkowski (1989). A photosynthetic quotient of unity was assumed in converting the O_2 evolution data of Osborne and Geider (1986) and Herzig and Falkowski (1989) to carbon assimilation rates.

from 0.0045 to 0.0065 $m^2 \cdot mg^{-1}$ chl a, but ϕ_m decreased from 0.129 to 0.10 mol $O_2 \cdot mol^{-1}$ photons. Herzig and Falkowski (1989) observed an increase in α^{chl} from 0.0094 to 0.016 $m^2 \cdot mg^{-1}$ chl$a$, and a decrease in ϕ_m from 0.08 to 0.06 mol $O_2 \cdot mol^{-1}$ photons. In contrast to these observations for nitrate-limited cultures, photon yields were approximately 3–4 times lower in vitamin B_{12}-limited *Monochrysis lutheri* than in nutrient-sufficient cultures. More dramatic declines of ϕ_m and α_{chl} have been observed in stationary phase batch cultures (Welschmeyer and Lorenzen, 1981; Cleveland and Perry, 1987), indicating a large reduction in photosynthetic efficiency under nitrogen-starved conditions.

Temperature dependence of photosynthesis

Most investigations of the temperature dependence of photosynthesis have examined the response of photosynthesis rates or enzyme activities

immediately following a shift in temperature from the growth temperature to a higher or lower experimental temperature. A theoretical framework for interpreting these transient responses is provided by the Arrhenius plot in which the natural logarithm of the rate constant of interest
is plotted against the inverse of the absolute temperature. The linear
portion of an Arrhenius plot fits the equation

$$\log_e k = -(E_a/RT) + C \qquad (7.21)$$

where k is the rate constant (s^{-1}), E_a is the activation energy ($kJ°/mol^{-1}$),
R is the universal gas constant ($8.3 \ mol^{-1} \cdot k^{-1}$), T is the absolute temperature (°K) and C is an arbitrary constant. Plots of ln k versus $1/T$ are linear
over a range of temperatures near the growth temperature but may show
a change in slope within this range. Outside of this temperature range,
k declines precipitously. Arrhenius constants for light-saturated photosynthesis and for various partial reactions have been summarized by
Raven and Geider (1989). Light-saturated photosynthesis and the carboxylase activity of RUBISCO are characterized by an E_a of 54–72 $kJ \cdot mol^{-1}$,
which corresponds to a Q_{10} of 2.1–2.7. This is slightly greater than the
Q_{10} for maximum (i.e., resource unlimited) growth (μ_m) of microalgae
obtained from interspecific comparisons, but lower than the Q_{10}, which
characterizes the phenotypic response of μ_m to temperature (Eppley,
1972). When considering a seasonal cycle, temperature is the environmental factor that consistently accounts for the largest component of the
variance in P_m^{chl} in marine phytoplankton assemblages with 50% of the
variation in P_m^{chl} accounted for by variations in temperature (Harrison and
Platt, 1980; Cote and Platt, 1983).

Recent investigations demonstrate that α_{chl} shows a similar temperature dependence to P_m^{chl}, immediately following a shift in temperature
(Priscu and Goldman, 1984; Tilzer et al., 1986; Palmisano et al., 1987).
This contrasts with the commonly held expectation that light-harvesting
and light-limited photosynthesis should be independent of temperature
(Talling, 1957a, b; Jewson, 1976). The temperature dependence of α^{chl} is
not limited to short periods following a change in temperature but can
be seen in samples from the natural environment, as well. For example,
Cote and Platt (1983) report that temperature can account for 56% of the
variance in α^{chl} within the temperature range 5–16°C in Bedford Basin,
Canada. This is similar to the proportion of variance in P_m^{chl}, which can
be attributed to temperature (Cote and Platt, 1983).

Reductions in α^{chl} at low temperature indicate that the maximum photon yield of photosynthesis (ϕ_m) may be temperature dependent, although it is not possible to confirm this suggestion without independent
measurements of the chlorophyll a-specific light absorption coefficient
(a^*). It is now well established that ϕ_m in terrestrial C_3 species is tempera-

ture sensitive because of the differential sensitivity of the carboxylase and oxygenase activities of the RUBISCO enzyme to variations in temperature (Ehleringer and Björkman, 1977). Observations that the light-limited growth rate is often reduced at low temperatures to a greater extent than would be expected on the basis of a reduction in pigmentation supports the suggestion that ϕ_m is reduced at low temperature (Raven and Geider, 1988). However, this is not consistent with an effect on RUBISCO kinetics as the oxygenase activity would be predicted to decrease at low temperatures resulting in an increase in ϕ_m (Ehleringer and Björkman, 1977). Other evidence indicating an effect of temperature on ϕ_m is the temperature dependence of growth yields (dry weight of biomass produced per mole of photons absorbed) of *Chlorella vulgaris* and *Chlorogonium* sp., although in the case of these organisms the temperature effect was attributed to changes in bulk biochemical composition and respiratory metabolism (Lee et al., 1985). There is a clear need for examining the temperature dependence of the photosynthetic energy conversion efficiency and algal growth, because many natural environments are characterized by temperatures considerably lower than the 15–25°C commonly employed in the laboratory.

Acclimation of an organism to a new temperature may modify the properties of catalysts and membranes, or change the relative amounts of different components and so influence photosynthesis rates in ways not predicted by short-term responses to temperature. The most striking acclimation to different temperatures involves reductions in chl*a*: C and chl*a* cell^{-1} at low temperatures (Geider, 1987; Davison, 1987), and increases in the activity of photosynthetic enzymes such as RUBISCO (Li and Morris, 1982; Davison, 1987). Davison (1987) demonstrated that light-saturated photosynthesis rates at an assay temperature of 15°C were linearly related to in vitro RUBISCO activities determined at a temperature of 20°C in *Laminaria saccharina* acclimated to temperatures between 0 and 20°C. He found identical activation energies (i.e., the slope of Arrhenius plot) and temperature dependencies of half saturation constants for TCO$_2$ uptake in plants previously acclimated to 5°C and 20°C.

Genetic Variability

Phenotypic variations in photosynthetic physiology and growth rate in response to environmental conditions were considered in the preceding sections. These phenotypic responses have often formed the basis for modeling growth of algal populations or assemblages. Good agreement, for instance, was found between the growth rate of natural populations of *Skeletonema costatum* in cage cultures and predictions based on the response of axenic cultures of a single strain of *S. costatum* to irradiance, temperature, and silicate concentration in the laboratory (Yoder, 1979). However, few rigorous comparisons of predictions based on laboratory experiments and measurements of growth or photosynthesis in nature have been undertaken, and the validity of extrapolating from observations on a relatively few species maintained in the laboratory to mixed assemblages growing under natural conditions has not been firmly established. One factor that might limit this autoecological approach is genetic variability, both between different clones of morphologically indistinguishable organisms and between different taxonomic species. Here we examine the magnitude of intraspecific (clonal) and interspecific genetic variability of algal responses to different environmental conditions. Unfortunately, despite the importance of assessing clonal and interspecific variations in algal physiology, there have been few investigations of the effects of genetic variation on algal photosynthesis. It has been necessary to supplement data on instantaneous photosynthetic responses with information on algal growth in order to address this problem more comprehensively.

Clonal Variability

Clonal variability in the growth rate of various marine phytoplankters has been most extensively examined in a series of investigations undertaken by Brand (Table 7.3). Brand (1981) found statistically significant differences in the growth rates of different clones isolated from the same water mass, and significant differences in population growth rates from different water masses. Different populations were characterized by coefficients of variation in reproduction rate, which ranged from 3 to 12% under controlled experimental conditions. Consistent with these observations, Krawiec (1982) found little clonal variability in maximum reproduction rates and temperature tolerance of six clones of *Thalassiosira rotula* isolated from Narragansett Bay. A single isolate from Baja, California had a significantly higher maximum division rate and a different temperature tolerance than the Narragansett Bay isolates. However, in both the Baja clone and a Narragansett Bay clone the assimilation number (P^{chl}) showed a similar dependence on temperature and irradiance. Gallagher (1982) found much higher variability in reproduction rates of *Skeletonema costatum* isolated from Narragansett Bay, with the clones under investigation having a coefficient of variation of reproduction rate of approximately 70%. However, Gallagher's (1982) cultures were not maintained in balanced growth, and the assessment of the significance of the observed clonal variability is difficult to evaluate. Although variations of up to

Table 7.3.   Genotypic (clonal) variations in division rate of various phytoplankters.

Species	Temp C°	Divisions/Day Range	c.v., %	n
Prorocentrum micans[1]	22	0.84–0.92	—	19
Gulf of Maine clone				
	26	0.65–1.1	—	19
Georges Bank clones	22	0.81–0.92	—	9
	26	0.96–1.11	—	9
Emiliania huxleyi[2]				
(Sargasso Sea clones from different locations)				
Population D	16	1.24–1.71	7.9	51
Population N	16	0.94–1.58	13.4	9
Population P	16	1.22–1.71	8.7	11
Population R	16	1.20–1.66	12.2	9
Population D	26	2.05–2.56	6.5	51
Population N	26	2.16–2.28	5.1	9
Population P	26	1.93–2.65	8.6	11
Population R	26	2.27–2.58	3.1	9
Gephyrocapsa oceanica[2]	16	0.89–1.22	9.0	9
(Sargasso Sea)				
Cyclococcolithina leptopora[2]	20	0.88–1.00	4.1	31
(Challenger Bank)				
	24	0.67–0.88	6.9	31
Prorocentrum micans[2]	22	0.84–0.92	3.2	21
(Gulf of Maine)				
	26	0.65–1.11	12.4	21
Dissodinium lunula[2]	24	0.29–0.37	6.6	41
(Sargasso Sea)				
Gonyaulax tamarensis[2]	16	0.19–0.66	10.4	57
(Falmouth, Massachusetts)				
Skeletonema costatum[3]	20	0.07–3.1	71	12
(Narragansett Bay, Rhode Island)				
Thalassiosira rotula[4]		2.3–2.4	2.1	6
(Narragansett Bay, Rhode Island)				

Data from (1) Brand (1981), (2) Brand (1985), (3) Gallagher (1982), and (4) Krawiec (1982).

10% in growth rate within a population are large by the yardstick of a geneticist, they are relatively small when compared with phenotypic variability.

Phenotypic and genotypic components of variability in the photosynthetic apparatus of *Skeletonema costatum* were investigated by Gallagher et al. (1984). Three clones that varied in maximum specific growth rate and electrophoretic mobility patterns were cultured under conditions of nutrient-sufficient balanced growth at both high and low irradiance.

Table 7.4.   Phenotypic and genotypic variability of the photosynthetic apparatus in *Skeletonema costatum*.

	Clone NY 17		Clone UP45		Clone Skel	
	27	274	27	274	27	274
10^{-15} mol chla:cell	0.38	0.30	0.74	0.57	0.67	0.40
mol chlc:mol chla	0.31	0.23	0.23	0.20	0.28	0.29
mol fuco:chla	0.52	0.40	0.48	0.29	0.20	0.09
mol chla:mol P_{700}	651	551	1380	951	558	516
mol chla:mol PSU(O_2)	579	416	610	494	478	327
α^{chl} [mol $O_2 \cdot mol^{-1}$ chl$_a$ $\cdot min^{-1}(\mu mol \cdot m^{-2} \cdot s^{-1})^{-1}$]	3.1	4.0	3.1	2.2	3.8	3.9
P_m^{chl} (mol $O_2 \cdot mol^{-1}$ chla$\cdot min^{-1}$)	5.8	7.7	6.2	4.7	9.4	9.9

Abstracted from Gallagher et al. (1984).

Significant genotypic and/or phenotypic differences were observed in a number of variables, some of which are summarized in Table 7.4. However, phenotypic variability of P_m^{chl} and α^{chl} showed no consistent pattern. Both P_m^{chl} and α^{chl} were independent of irradiance in one clone, but were positively related to irradiance in a second clone and inversely related to irradiance in a third. However, covariation of α^{chl} and P_m^{chl} in these clones of *S. costatum* may indicate that changes in the photosynthetic apparatus are constrained by similar underlying physiological processes. A similar covariation of α_{cell} and P_m^{cell} was found in two *Synechococcus* clones cultured under a range of conditions (Glibert et al., 1986).

Ecotypic and developmental differences in the photosynthetic characteristics of *Laminaria saccharina* were investigated by Gerard (1988) with plants taken from shallow- and deep-water sites in Maine and a turbid site in New York State. Juvenile *L. saccharina* were collected from the field sites and grown 6 weeks at four different light levels (0.065, 0.12, 0.26, and 0.54 of the incident irradiance), with nitrate and phosphate added at 2-d intervals. Gerard (1988) found that area-based measurements of P_m and α were proportional to chlorophyll *a* content, but that plants varied in chlorophyll content, such that the turbid-water ecotype outperformed both the shallow-water and the deep-water forms at all irradiances. Plants from the deep-water site had lower specific growth rates under both light-saturated and light-limited conditions than plants from the shallow-water site, and plants from the turbid site had the highest growth rates at all irradiances. Significantly, P_m^{chl}, α^{chl}, and the photon yield of photosynthesis were similar in all three ecotypes. Thus, ecotypic variation in *Laminaria saccharina* was due primarily to differences in pigment content rather than to fundamental differences in the chlorophyll-specific rate of photosynthesis.

Interspecific variability

Examining interspecific variability in photosynthetic gas exchange rates is complicated by the phenotypic and clonal variability outlined above. However, when organisms are cultured under identical experimental conditions, it is possible to examine the variability in specific growth rates, photosynthesis rates, or biochemical composition, which can be attributed to genetic variation. When such comparisons are made, much of the interspecific variability can be accounted for by higher taxonomic affiliation and organism size in the microalgae (Banse, 1982) and by taxonomic affiliation, size, and form in the macroalgae (Littler and Arnold, 1982; Littler et al., 1983). However, there must be a careful consideration of the appropriate physiologic attributes used in any comparison. For example, Dunstan (1973) found that five species of chlorophyll c-containing microalgae could be characterized by identical values of I_k, which implies little interspecific variability, but that chlorophyll a-specific light-saturated photosynthesis and dark respiration rates varied by about a factor of two, implying rather more variability. This section begins by considering the magnitude of interspecific variability, before proceeding to a discussion of the extent to which interspecific variability is predictable.

Maximum, resource-unlimited growth rates (μ_m) of microalgae, growing under optimal conditions, vary by over an order of magnitude. For example, at 20°C, μ_m can be expected to range from approximately 0.1 d^{-1} in some large slow-growing dinoflagellates to approximately 3 d^{-1} in some small fast-growing diatoms and green algae. This 30-fold range may approximate an upper limit on the amount of interspecific variability in μ_m that will be observed under controlled conditions. In fact, the range in μ_m is less than fivefold in many interspecific comparisons (Table 7.5), even when organisms have been deliberately chosen to include both fast- and slow-growing species.

To make meaningful assessments of interspecific variability in growth and photosynthesis, it is necessary to restrict the scope of the comparison to well-defined higher taxonomic groups. Table 7.5, for example, examines the variability in growth rates within the cyanobacteria, diatoms, dinoflagellates, and green algae. For comparison, Table 7.6 examines the variability of some indicators of the size (C:chl$a$, chl$a$:protein) or operating performance (P^{chl}, P_m^{chl}, α^{chl}) of the photosynthetic apparatus. This information is limited to those studies in which four or more species or clones were cultured under relatively well-defined environmental conditions. Within each of the taxa under consideration, μ_m varied by up to a factor of five and coefficients of variation for μ_m varied by up to 140%. This interspecific variation is considerably greater than that observed for

Table 7.5. Interspecific variability of growth rate within higher taxonomic groups of marine and freshwater phytoplankton.

	Light:Dark Period	T, °C	E	Max/Min	c.v., %	n
Cyanobacteria[1]	C*	20	60	5.1	38	22
Diatoms[2]	C	18	320	3.5	49	6
Diatoms[3]	C	21	8	1.6	19	5
	C	21	16	1.5	16	5
	C	21	32	1.2	8	5
	C	21	80	1.4	13	5
	C	21	160	1.4	15	5
	C	21	256	4.9	43	5
Diatoms[4]	C	24	35	—	84	10
	C	24	80	—	79	9
	C	24	350	—	96	10
	C	24	800	—	135	10
	14:10	24	35	1.9	18	10
	14:10	24	80	1.9	19	9
	14:10	24	350	2.5	27	10
	14:10	24	800	2.9	33	10
Dinoflagellates[3]	C	21	8	4.2	54	5
	C	21	16	2.4	42	5
	C	21	32	3.8	46	5
	C	21	80	3.9	45	5
	C	21	160	—	79	5
	C	21	256	—	120	5
Dinoflagellates[3]	C	24	35	—	87	6
	C	24	80	—	64	
	C	24	350	—	84	7
	C	24	800	—	125	7
	14:10	24	35	—	79	
	14:10	24	80	6.8	53	6
	14:10	24	350	—	65	7
	14:10	24	800	—	115	6
Coccolithophorids[4]	C	24	35	—	67	4
	C	24	80	—	68	4
	C	24	350	—	70	4
	C	24	800	—	70	4
	14:10	24	35	2.3	32	4
	14:10	24	80	2.0	28	4
	14:10	24	350	2.8	38	4
	14:10	24	800	3.3	51	4
Green algae[5]	C	16	10	1.2	11	4
	C	16	25	2.0	26	4
	C	16	40	2.8	42	4
	C	16	60	2.4	39	4
Green Algae[5]	C	26	10	1.7	22	4
	C	26	25	2.5	43	4
	C	26	40	2.6	53	4
	C	26	60	4.8	54	4

Data obtained from (1) Foy (1980), (2) Blasco et al. (1982), (3) Chan (1978), (4) Brand and Guillard (1981), and (5) Schlesinger et al. (1981).
*C denotes continuous light.

Table 7.6.   Interspecific variability in photosynthesis and chlorophyll a: biomass ratios within higher taxonomic groupings of marine and freshwater phytoplankton.

Taxonomic group	T, °C	E, $\mu mol \cdot m^{-2} \cdot s^{-1}$	Max/Min	c.v., %	n
Cyanobacteria[1]					
P_m^{chl}	20	30	4.7	27	20
P_m^{chl}	20	150	2.3	27	20
α^{chl}	20	30	1.7	18	20
α^{chl}	20	150	1.9	20	20
Diatoms[2]					
P_{chl}	18	320	5.8	52	6
C:chl$a$	18	320	1.7	22	6
C:N	18	320	1.4	13	6
Diatoms[3]					
P_m^c	5–8	50	52.0	142	16
α^{chl}	5–8	50	13.0	65	16
C:chl$a$	5–8	50	4.5	54	16
Diatoms[4]					
chl$a$:protein	21	8	2.0	27	5
chl$a$:protein	21	16	1.6	18	5
chl$a$:protein	21	32	1.4	12	5
chl$a$:protein	21	80	1.3	18	5
chl$a$:protein	21	160	2.2	31	5
chl$a$:protein	21	256	3.6	43	4
Dinoflagellates[4]					
chl$a$:protein	21	8	1.5	25	5
chl$a$:protein	21	16	2.4	45	5
chl$a$:protein	21	32	2.4	28	5
chl$a$:protein	21	80	2.1	29	4
chl$a$:protein	21	160	—	70	5
chl$a$:protein	21	256	—	104	5

Data from (1) Foy and Gibson (1982a), (2) Blasco et al. (1982), (3) Chan (1978), and (4) Taguchi (1976).

clonal variability in which coefficients of variation of approximately 10% are typically observed (Table 7.3). Photosynthetic characteristics are considerably less variable than μ_m, with coefficients of variation of 25 to 50% for P_m^{chl} and α^{chl}, and 10 to 40% for the C:chl$a$ and protein: chl$a$ ratios. Apparently exceptional in this regard are Taguchi's (1976) observations in which C:chl$a$, P_m^c and α^{chl} vary widely.

Higher taxonomic affiliations can account for a large fraction of the interspecific variability noted above. It is commonly observed that dinoflagellates grow slower than diatoms when cultured under identical environmental conditions (Chan, 1978; Brand and Guillard, 1981), whereas chlorophytes (Schlesinger and Shuter, 1981; Schlesinger et al., 1981) and coccolithophorids (Brand and Guillard, 1981) appear, on average, to grow at rates comparable to those achieved by the diatoms.

Organism size also appears to account for a significant fraction of interspecific variability. Generally the rates of metabolic processes in both unicellular and multicellular organisms have size dependencies that can be described by an allometric relation of the form

$$rate = aW^b \tag{7.22}$$

where *rate* is the rate of some metabolic process, W is a measure of organism size, and a and b are constants (Peters, 1983). When specific metabolic rates with units of inverse time are plotted against mass, the exponent of Eq. 7.16 usually varies from -0.25 to -0.33 (Fenchel, 1974; Peters, 1983). In other words, specific metabolic rates decrease with increasing organism size. Within the microalgae, the allometric dependence only becomes evident when organisms from a wide range of cell size are considered. Groups of organisms that meet these criteria and for which sufficient data is available to examine the size dependence of metabolic rates, include the diatoms (Banse, 1976, 1982; Chan, 1978; Blasco et al., 1982; Geider et al., 1986), dinoflagellates (Chan, 1978, 1980; Banse, 1982), chlorophytes (Schlesinger and Shuter, 1981; Schlesinger et al., 1981), and cyanobacteria (Foy, 1980; Foy and Gibson, 1982a).

Growth rate is one of the few variables for which a size dependence has been examined in microalgae. When data from all classes of microalgae are included in allometric equations, growth rate scales approximately with mass as $W^{-0.30}$ (Schlesinger et al., 1981; Langdon, 1988), apparently consistent with observations for other groups of organisms. However, when data are grouped by higher taxonomic affiliation, the size dependence is markedly reduced (Banse, 1982) as indicated in Table 7.7. Schlesinger et al. (1981) provide evidence for a reduced size dependence of growth under suboptimal environmental conditions previously postulated by Banse (1976). Significantly, organism size can account for approximately 75% of the variance in growth rate (Table 7.7).

Bulk biochemical composition shows little size dependence within particular classes of microalgae (Chan, 1978; Hitchcock, 1982; Schlesinger et al., 1981), although it may vary significantly between higher taxonomic groups. In particular, dinoflagellates are characterized by lower chlorophyll a:biomass ratio than diatoms (Geider, 1987), but it is unclear how relative pigment concentrations vary among the other algal classes or in the cyanobacteria. There is some evidence for a reduction in the ratio of chlorophyll a to biomass as cell size increases (Chan, 1978; Blasco et al.,

Table 7.7.   Size dependence of growth in microalgae.

	T, °C	μ, d^{-1}	r^2	n
All taxa included[1]	20	$3.51C^{-0.32}$	0.70	26
All taxa included[2]	15	$2.95C^{-0.28}$	0.59	32
Diatoms[3]	18	$2.60C^{-0.14}$	0.91	56
Diatoms[4]	21	$2.28C^{-0.06}$		5
Diatoms[5]	20	$3.02C^{-0.13}$	0.66	10
Dinoflagellates[4]	21	$1.61C^{-0.16}$		5
Dinoflagellates[5]	21	$1.38C^{-0.15}$	0.75	6
Chlorophytes[6]	16	$1.09C^{-0.19}$	0.72	4

Data from (1) Schlesinger et al. (1981) based on a literature review, (2) Langdon (1988), based on a literature review, (3) Blasco et al. (1982), (4) Chan (1978) assuming that pg protein/cell = pg C/cell, (5) Banse (1982) (note that Banse's regression for dinoflagellates is based largely on Chan's 1978 data), and (6) Schlesinger et al. (1981) data for an irradiance of 12 W·m^{-2}.

1982), but this reduction in pigmentation does not appear to be sufficient to offset a reduction in chlorophyll a-specific photosynthesis rates (P^{chl}) due to a reduced growth rate (Blasco et al., 1982).

The size-dependence of growth rate discussed above should not be allowed to overshadow the potential use of other functional attributes as predictors of algal growth and photosynthesis, especially in the light of the weak size dependence often observed for specific growth rate. It is well established that a significant proportion of phenotypic variation in growth can be accounted for by differences in the chlorophyll a-to-carbon ratio and irradiance (Kiefer and Mitchell, 1983; Sakshaug et al., 1989). However, the use of variables such as the chlorophyll a-to-carbon ratio to describe interspecific variations in μ are less well established. An important exception is the observation by Chan (1978) that most of the variation in nutrient-sufficient specific growth rates of diatoms and dinoflagellates could be accounted for by differences in irradiance and the chlorophyll a-to-protein ratio. These results indicate that algal growth rates can be predicted by considering the combined effects of one environmental variable and one physiological variable. Specifically, growth rate was found to be linearly related to the chlorophyll a-to-protein ratio at each of the six irradiances examined by Chan (1978). Underlying this relationship between μ, and the chlorophyll a-to-protein ratio is an implicit limitation on the interspecific variability in the parameters P_m^{chl} and α^{chl} of the photosynthesis–light response curve (Chan, 1980).

Problems in Measuring Parameters of the Photosynthesis–Light Response Curve

Although often considered to be an attribute solely of the organism being examined under a given set of environmental conditions, the PI curve is not fixed exclusively by these factors, but also depends in a complex manner on the duration of the experiment. Two extremes in duration can be imagined. At one extreme, the response of growth rate to irradiance is obtained after exposure to continuous, usually constant irradiance, for a period of days to weeks. At the other extreme, the response of the water-splitting reaction of PS2 to irradiance can be obtained through flash yield experiments in which individual flashes of < 1-μs duration are used to probe the operation of PS2. It is to be expected that growth rate and PS2 photochemistry will have different dependencies on irradiance due to differing rate limitations and regulatory activities, although direct comparisons of the light-response curves for PS2 activity and growth have not been undertaken.

The appropriateness of current protocols: a question of time scale

Most determinations of the PI curve employ experimental durations of minutes to hours, a range of time scales that is clearly intermediate to the two extremes mentioned above. Transients in the O_2 evolution rate that accompany the onset of illumination, or a large change in irradiance, are well characterized and are typically complete within a period of seconds. The steady rate of O_2 evolution over a period of minutes that follows these transients is often used in determinations of the PI curve. Myers (1970) was the first to discuss the difference between the photosynthesis–light response curves obtained from these nearly instantaneous observations and the longer term dependence of growth rate on irradiance. When expressed in the appropriate units, response curves for growth and photosynthesis are expected to cross at the irradiance to which the algae are acclimated. Comparisons of PI and growth irradiance response curves are difficult to make because of differences in the geometries of culture vessels, light sources, and methods employed to measure growth and photosynthesis. A recent comparison for *Phaeodactylum tricornutum* is illustrated in Fig. 7.7, where it can be observed that the instantaneous photosynthesis rate exceeds the growth rate at irradiances above that to which the organism was acclimated, but that the growth rate exceeds the photosynthesis rate at lower irradiances.

Even when PI curves are obtained in short-duration experiments, it is found that the rate of photosynthesis at a given irradiance may depend on the previous irradiance (Marra et al., 1985; Neale and Marra, 1985; Dromgoole, 1987). Harris (1973) and Harris and Piccinin (1977) called attention to the problem that instantaneous rates of photosynthesis observed in rising light may exceed rates observed in falling light, and termed the divergence in measured rates of photosynthesis "hysteresis." Harris (1973) noted that hysteresis was more pronounced in samples

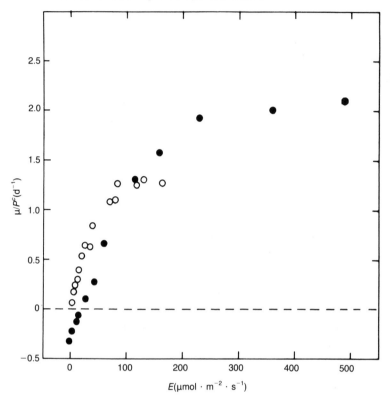

Fig. 7.7. Comparison of the photosynthesis-irradiance and growth-irradiance curves of the marine diatom *Phaeodactylum tricornutum* (Geider and Osborne, unpublished). Open circles indicate the irradiance dependence of growth rate (μ, in units of d^{-1}), whereas closed circles indicate the irradiance dependence of the carbon-specific photosynthesis rates (P^c, in units of d^{-1}), for cells grown at an irradiance of 100 μmol photons·m^{-1}·s^{-1}. Note that the two curves cross at the growth irradiance, and that P^c is less than μ at irradiance less than the growth irradiance, whereas P^c exceeds μ at irradiances above the growth irradiance.

exposed to saturating and photoinhibiting irradiances than in samples exposed to subsaturating irradiances, and that the magnitude of hysteresis appeared to depend on the duration of exposure to high irradiances. Hysteresis in net gas exchange measurements has been attributed to several phenomena, including photoinhibition, photorespiration, and variable dark respiration (Harris, 1978; Falkowski et al., 1985). It is possible that all of these processes contribute to hysteresis to a variable extent depending on the organism under investigation and the experimental conditions.

Clearly, the shape of the PI curve and the magnitude of the parameters that describe this curve may be affected by hysteresis, should it occur to any significant extent. To standardize procedures employed to measure the PI curve, many investigators begin by determining the oxygen consumption rate in darkness followed by measurements of net O_2 evolution at increasing irradiances. This approach should minimize the possibility that photoinhibition will affect the determination of the light-limited photosynthesis rates. Myers and Graham (1971) recommend that samples be directly transferred from the growth chamber to the cuvette for measurement of gas exchange rate within 10 min of removal from the previous growth condition. This avoids any influence of previous treatments on measured photosynthesis rates, but the precision of the analysis may suffer since measurements of oxygen exchange are made on separate samples taken at separate times. Determination of the PI curve using ^{14}C-bicarbonate assimilation employs parallel incubation of a set of subsamples at a range of irradiances. Thus, each subsample is exposed to a different irradiance for the same time interval, which may range from minutes to hours. Sample size can vary from 1 cm^{-3} incubated directly in a scintillation vial (Lewis and Smith, 1983), to a more typical volume of 50 to 200 cm^{-3} (Jassby and Platt, 1976).

The preceding discussion indicates that the photosynthetic response of algae may not always be unambiguously described by PI curves. The utility of the PI curve decreases for algae exposed to very high irradiances for an extended period of time. In particular, photosynthesis rates decline during long exposure to photoinhibiting irradiances, with both the initial rate and the rate of decline dependent on irradiance (Kok, 1956; Marra, 1978, 1980). The rate of decline and any subsequent recovery depends not only on the experimental irradiance, but also on the physiological condition of the algae under investigation. As a consequence, the parameters of the PI curve are seen to depend on the duration of the experiment as well as on the characteristics of the sample.

Chapter 8

Remote Sensing of Algal Photosynthesis

Estimates of the rates of photosynthetic energy conversion on regional and global scales are required for predicting the energy available to food chains and for determining the role of algae in geochemical cycles (Sarmiento and Toggweiler, 1984). As all organisms are ultimately dependent on energy input *via* photosynthesis, and limitations to global carbon dioxide assimilation may constrain the potential productivity of higher trophic levels (Ryther, 1969). Riley (1944) provided the first estimate of global oceanic primary productivity of $860 \pm 560 \, mg \, C \cdot m^{-2} \cdot d^{-1}$. The large standard deviation associated with this estimate arose from uncertainty in extrapolating a small number of observations from a limited range of open and coastal ocean regions to the entire ocean. Despite an additional 45 years of research involving the widespread incorporation of ^{14}C assimilation measurements into biological oceanographic sampling programs, there is continuing uncertainty about the magnitude of oceanic primary productivity (National Academy of Sciences, 1984). Shipboard observers cannot adequately sample the ocean at all spatial and temporal scales necessary to adequately resolve variations in phytoplankton biomass and productivity. In contrast, satellite remote sensing offers the potential to provide a detailed assessment of phytoplankton spatial and temporal variations. The remotely sensed signal is used to infer chlorophyll *a* concentration from measurements of light reflected from near surface waters. At present, remote sensors do not provide a direct measure of primary productivity; this must be inferred from the relationship between primary productivity and chlorophyll *a* concentration.

It has long been a challenge to estimate phytoplankton productivity from chlorophyll *a* concentration, because pigment concentration can be sampled much more rapidly than photosynthesis (Ryther and Yentsch,

1957). Introduction of in vivo fluorometry to continuously monitor chlorophyll concentration (Lorenzen, 1966), together with radiometric sensors for estimating chlorophyll a from aircraft or satellites (Gordon and Morel, 1983) and moored fluorometers to continuously measure chlorophyll a at fixed positions in continental shelf waters (Whitledge and Wirick, 1986), have greatly increased the spatial and temporal resolution with which phytoplankton abundance can be determined. Unfortunately, there has not been a corresponding increase in the ability to measure phytoplankton photosynthesis, even though deployment of high-resolution O_2 sensors (Maccio and Langdon, 1988), pump and probe fluorometers (Kolber et al., 1990), and sensors to measure passive chlorophyll fluorescence (Kiefer et al., 1989) may allow increases in the temporal and spatial coverage of phytoplankton primary production. In lieu of more widespread application of these techniques, investigators have attempted to estimate primary production from chlorophyll a concentration and a range of environmental variables, using empirical and analytical algorithms. This chapter considers (1) the methods for remote sensing of the chlorophyll a concentration in surface waters and (2) the ways in which chlorophyll a distributions can be used to estimate primary production.

Remote sensing of phytoplankton pigment concentration can be achieved by (1) passive measurements of water color (Gordon et al., 1983), (2) passive measurements of solar-induced chlorophyll fluorescence (Neville and Gower, 1977), and (3) active measurement of laser-induced fluorescence (Hoge and Swift, 1981; Sathyendranath, 1986). Water color is quantified by measuring the radiance upwelling from the surface waters using shipboard, aircraft, or satellite radiometers. Satellite instruments currently employ radiometers with broad spectral bands (60–100 nm) in the visible region, limiting measurements to a single pigment (chlorophyll a) in clear waters, with a low diversity of optically active components. However, in clear ocean waters, the ocean color algorithms for remote sensing of chlorophyll a yield relatively good precision over a wide range of chlorophyll a concentrations without the need for extensive, independent ground truth observations. Much greater spectral resolution is possible with ship and aircraft borne radiometers (Carder and Steward, 1985), with the potential for evaluating the concentrations of several major pigments even in turbid coastal waters with a large diversity of optically active components. However, accurate remote sensing of the phytoplankton concentration in a complex mixture first requires more detailed knowledge of the optical properties of phytoplankton (Chapter 6), as well as measurements of the other components of these assemblages, such as particulate material and colored substances (Carder and Steward, 1985). This information becomes more directly applicable to satellite remote sensing during the 1990s with the development and deployment of satellites that will carry moderate- and high-resolution spectrometers.

Passive remote sensing of in vivo chlorophyll *a* fluorescence relies on high spectral resolution measurements of upwelling radiance at wavelengths around 680 nm (Gower, 1980; Sathyendranath, 1986; Topliss and Platt, 1986). The technique is based on the same principals as in vivo fluorometry (Chapter 3), but differs from the in vivo technique in two important respects. First, passive remote sensing of chlorophyll *a* fluorescence relies on a natural, variable light source (the sun) instead of an artificial source with a known spectral quality and irradiance. Second, remote sensing of chlorophyll *a* fluorescence samples the upwelling irradiance that leaves the water surface rather than a known small volume. The remotely sensed fluorescence emission is, however, superimposed on a backscatter signal of unknown magnitude (see Chapter 6), which can complicate the interpretation of the fluorescence signal. In addition, the approximately order-of-magnitude variability of in situ fluorescence yield (Carder and Steward, 1985; Topliss and Platt, 1986), for reasons discussed in Chapter 3 and 5, necessitates ground truth observations for calibration purposes.

Active remote sensing of surface phytoplankton abundance relies on the use of lasers pulsed at short intervals to excite in vivo chlorophyll *a* fluorescence and a photomultiplier to measure the induced red (680 nm) fluorescence (Yentsch and Yentesch, 1984). The primary fluoroscensor in use today is the Airborne Oceanographic Light Detection and Ranging Detector (AOL) developed at the NASA Goddard Wallops Island Flight Station (Hoge and Swift, 1981; Hoge et al., 1986a). In this system, the excitation wavelength is 532 nm and the return signal to an airborne spectrometer includes contributions from backscattering at the excitation wavelength, Raman scattering by water, and fluorescence by chlorophyll *a* and phycoerythrin (Exton et al., 1983; Hoge et al., 1986b). Laser systems for measuring pigment concentrations are limited to low-flying aircraft largely because of the power requirements for operating the laser, atmospheric attenuation of the signal, and bulky instrumentation (Sathyendranath, 1986). The AOL allows simultaneous measurement of active laser-induced chlorophyll fluorescence, passive solar-induced chlorophyll fluorescence, and passive ocean color (Hoge et al., 1986a, b), and can be used to obtain simultaneous estimates of chlorophyll concentrations from passive and active techniques (Hoge et al., 1986a, b). Good correlations between laser-induced chlorophyll *a* fluorescence and ground-truthed chlorophyll *a* concentrations suggest that variations in fluorescence yield are not significant during the approximately 1–2-hr flight time typically employed in experiments using the AOL.

Estimating Surface Water Chlorophyll *a*

Most phytoplankton abundance measurements to date have relied on ocean color measurements using the now defunct Coastal Zone Color Scanner (CZCS) on the Nimbus 7 satellite. Estimating chlorophyll *a* from ocean color measurements rests on the observation that dissolved or suspended materials change the reflectance of water through the absorption and scattering of light. Radiative transfer theory serves as the basis for examining the relations among reflectance, absorption, and scattering (Gordon et al., 1975; Morel and Prieur, 1977). A model developed by Gordon et al. (1975) leads to the following equation relating subsurface reflectance to the absorption and back-scattering coefficients:

$$R_s(\lambda) = 0.33 \, b_b(\lambda) / [a(\lambda) + b_b(\lambda)] \qquad (8.1)$$

where $R_s(\lambda)$ is the subsurface reflectance (dimensionless), $a(\lambda)$ is the absorption coefficient (m^{-1}) and $b_b(\lambda)$ is the back-scatter coefficient (m^{-1}) (Table 8.1). The absorption and back-scatter coefficients with units of inverse length (i.e., m^{-1}) can in turn be expressed in terms of the concentrations and specific absorption or scattering coefficients of optically active material. It is commonly assumed for open ocean waters that most of the back scattering is due to water itself and most of the variable absorption is due to phytoplankton pigments together with substances that covary with phytoplankton abundance, leading to the approximation

$$R_s(\lambda) = 0.33 b_{bw}(\lambda) / [a^*(\lambda)C + a_w(\lambda) + b_{bw}(\lambda)] \qquad (8.2)$$

where $a^*(\lambda)$ is the typical chlorophyll *a*-specific absorption coefficient for phytoplankton ($m^2 \cdot g^{-1}$ chl*a*) and C is the chlorophyll *a* concentration ($g \cdot m^{-3}$). More complicated expressions, including contributions of dissolved and suspended inorganic matter or dissolved organic substances, such as humic and fulvic acids, are required for near-shore waters and lakes. Changes in the concentration of chlorophyll *a* will markedly affect reflectance in those spectral regions in which phytoplankton absorb light most strongly (i.e., in the blue and red). Algorithms for determining chlorophyll *a* concentration from CZCS images depend on an empirical relationship between the ratio of blue to yellow light upwelled through the ocean surface and chlorophyll *a* concentration. This relationship depends on the observation that the "typical" phytoplankton absorption spectrum has peaks centered in the blue ($\lambda = 440$ nm) and red ($\lambda = 670$

Table 8.1.   Summary of terms used in describing the relations between photosynthesis and phytoplankton concentration.

Symbol	Definition	Typical Units
C_a	Surface chlorophyll a concentration	g chla·m^{-3}
C_k	Mean chlorophyll a concentration within the top optical depth of the water column	g chla·m^{-3}
C_i	Integrated chlorophyll a content of the euphotic zone	g chla·m^{-2}
E_0	Incident scalar irradiance	mol photons·m^{-2}d^{-1}
I_k	Light saturation parameter	mol photons·m^{-2}d^{-1}
K	Vertical attenuation coefficient (log$_e$ units)	m^{-1}
P_m^{chl}	Chlorophyll a-specific light-saturated photosynthesis rate	g C·g^{-1}chla·d^{-1}
D	Depth of the euphotic zone defined as the depth at which irradiance equals 1% of surface value	m
π	Integrated euphotic zone primary productivity	g C·m^{-2}·d^{-1}
π_{chl}	Ratio of integrated primary productivity to integrated chlorophyll a ($\pi_{chl} = \pi/C_i$)	g C·g^{-1} chla·m^{-1}
Ψ	Light utilization efficiency defined as π_{chl}/E_0	g C·mol^{-1}photons·m^{-2}g^{-1}chla
α_{chl}	Chlorophyll a-specific light absorption coefficient	m^2·g^{-1}·chla
a_{chl}	Chlorophyll a-specific initial slope of the photosynthesis-light curve	g C·mol^{-1}photons·m^{2}g^{-1}chla
η	Realized photon yield for integral photosynthesis	g C·mol^{-1}photons
Φ_m	Maximum photon yield of photosynthesis	g C·mol^{-1}photons

nm) regions but does not absorb strongly in the yellow (λ = 550 nm). The algorithm has the form

$$\log [C] = a + b\log [R(443) / R(550)] \qquad (8.3)$$

where $R(\lambda)$ is the reflectance centered at wavelength λ and a and b are fitted coefficients obtained from a comparison of surface reflectance derived from satellite measurements corrected for atmospheric effects and sea surface chlorophyll a concentrations determined from ship observations (Gordon et al., 1983).

The approach (Eq. 8.3) is quasiempiric, resting on the observation that high absorption of blue light by phytoplankton reduces the reflectance in this spectral region relative to wavelengths in the yellow where pigment absorption is minimal. The ratio of reflectances in the green ($\lambda = 520$ nm) to yellow ($\lambda = 550$ nm) spectral bands of the CZCS have also been used to replace $R(443)/R(550)$ in Eq. 8.3 with equally good precision (Gordon et al., 1983; Holligan et al., 1983). Estimates of remotely sensed chlorophyll a are within $\pm 40\%$ of surface observations over a range of approximately 0.1–50 μg chl$a \cdot$dm^{-3} (Gordon et al., 1983; Holligan et al., 1983; Pan et al., 1988). Limits on the precision are due to errors in making atmospheric corrections, variability in phytoplankton optical properties among diverse phytoplankton assemblages, and variability in the optical properties of seawater due to the nonuniform distribution of dissolved organic matter and suspended sediments (Gordon et al., 1983; Holligan et al., 1983; Sathyendranath, 1986; Balch et al., 1989).

The reflectance ratio algorithms were developed for use in optically clear (Case 1) ocean waters in which chlorophyll a and substances that covary with chlorophyll a are the major sources of optical variability. Use of algorithms based on blue to green reflectance ratios are unlikely to resolve phytoplankton abundance in estuarine waters because of light absorption in the blue and green spectral regions by iron oxides and humic-acid-like material, which can be important in estuarine and freshwaters (Stumpf and Tyler, 1988). To overcome this problem, the ratio of reflectances in the red to near infrared, where absorption by dissolved materials is minimal, can be employed. The principal is the same as that employed by Gordon et al. (1983) for clear open ocean water measurements. High phytoplankton absorption in the red reduces reflectance at this wavelength relative to a region of low pigment absorption in the near infrared. This approach has been used by Stumpf and Tyler (1988) to measure chlorophyll a concentrations in excess of 5 μg chl$a \cdot$dm^{-3} in turbid estuarine waters with a precision of $\pm 60\%$ or 5 μg chl$a \cdot$dm^{-3}, whichever is the greater.

Although satellite remote sensing provides a means of greatly increasing spatial coverage, the technique is subject to a number of limitations including (1) restricted temporal coverage due to cloud cover, (2) inaccuracies in processing algorithms for making atmospheric corrections, particularly at large zenith angles (Pan et al., 1988), (3) poor vertical resolution that is limited to the upper 20% of the euphotic zone (Smith, 1981), which includes a small and variable (1% to 8%) portion of chlorophyll a in ocean waters (Platt and Herman, 1983). In addition, the depth interval that is included in a remotely sensed measurement will vary with wavelength (λ), being inversely proportional to the attenuation coefficient $K(\lambda)$. For example, as the chlorophyll concentration rises from 0.01 to 10 mg$\cdot$m^{-3}, the depth zone that accounts for 90% of the upwelling irradiance decreases from 58 to 2 m at 443 nm and decreases from 15 to 4 m at 550 nm (Sathyendranath, 1986), while remaining virtually constant at 2 m for 670 nm light (Smith, 1981). Thus, passive and active remote sensing of chlorophyll a fluorescence is limited to the very shallow depths near the surface, whereas observations of ocean color depend on light that penetrates deeper into the water column.

Other variables, in addition to chlorophyll a concentration, that are important in determining photosynthesis rates, and that can be measured by aircraft and/or satellite sensors include temperature and downwelling irradiance at the earth's surface (Gauthier and Katsaros, 1984) and the depth of the surface ocean mixed layer (Yan et al., 1991). These measurements will not be discussed further, but the importance of including these variables in algorithms for the remote sensing of phytoplankton photosynthesis will become apparent in the remainder of this chapter.

Empirical Relationships Between Primary Production and Chlorophyll *a*

As a first approximation, primary productivity is expected to be proportional to plant biomass, *all other factors being equal*. Both limnologists and oceanographers have observed linear relations between primary production and chlorophyll a concentration for data expressed on either a unit volume or unit area basis, with biomass accounting for 50% to 75% of the variance in primary production (Table 8.2) (Fee, 1979; Smith, 1979; Hayward and Venrick, 1982). For remote sensing applications, it is desirable to relate integral primary production (with units of g $C \cdot m^{-2} \cdot d^{-1}$ and designated π) to near-surface chlorophyll a concentrations (with units of mg chl$a \cdot m^{-3}$, and designated C_s). Despite the linear relation that is sometimes observed between π and C_s (Table 8.2), theoretical analyses suggest that the relationship between integral primary production and phytoplankton abundance should be nonlinear largely because of the influence of other light-absorbing substances (Talling, 1957a, b; Bannister, 1979; Platt, 1986). In fact, a logarithmic relation is often observed between areal primary production and surface phytoplankton abundance:

$$\log_e(\pi) = \log_e(a) + b \log_e (C_s) \tag{8.4}$$

or

$$\pi = aC_s^b \tag{8.5}$$

where a and b are empirically derived coefficients. Observations for various ocean regions are characterized by $b = 0.4\text{--}0.6$ (Table 8.3). In

Table 8.2. Empirical relations between primary production and chlorophyll concentration in lakes and selected marine environments.

Location	Dependent Variable	Independent Variable	Slope	Intercept	R^2
North Western Ontario[1] Experimental Lakes District (2–50 mg chla·m^{-2})	g C·m^{-3}·y^{-1}	g chla·m^{-3}	1900	−0.5	—
North Western Ontario[1] Experimental Lakes District (1–6 mg chla·m^{-3})	g C·m^{-3}·y^{-1}	g chla·m^{-3}	166	0.53	0.76
North American Lakes[2]	g C·m^{-3}·d^{-1}	g chla·m^{-3}	22.9	−0.0426	0.81
Southern California Bight[3]	g C·m^{-2}·d^{-1}	g chla·m^{-3}	0.35	—	0.33
California Current[4]	g C·m^{-2}·h^{-1}	g chla·m^{-2}	2.7	—	0.83
North Pacific Subtropical Gyre[4]	g C·m^{-2}·h^{-1}	g chla·m^{-2}	2.3	—	0.50
MERL Microcosm[5]	g C·m^{-2}·h^{-1}	g chla·m^{-2}	3.4(0.2)	0.478	0.51
MERL Microcosm[5]	g C·m^{-2}·d^{-1}	g chla·m^{-2}	34.2(2.4)	0.437	0.47
MERL Microcosm (Summer)[6]	g C·m^{-2}·d^{-1}	g chla·m^{-3}	125	0.25	0.76
MERL Microcosm (Nonsummer)[6]	g C·m^{-2}·d^{-1}	g chla·m^{-3}	53	0.12	0.36

Data from (1) Fee (1979), (2) Smith (1979), (3) Smith et al. (1982), (4) Hayward and Venrick (1982), (5) Keller (1988a), and (6) Keller (1988b).

Table 8.3.   Empirical relations between integral primary productivity and surface chlorophyll *a* concentration.

Empiric relationship	Primary productivity (g $C \cdot m^{-2} \cdot d^{-1}$)	Surface chlorophyll concentration (mg $chla \cdot m^{-3}$)	Correlation coefficient	Reference
$\pi = 1.53\ C_s^{0.457}$	0.06–11	0.04–28	0.539	Lorenzen (1970)
$\pi = 1.00\ C_s^{0.500}$	0.10–10	0.05–100	—	Eppley et al. (1985)
$\pi = 1.88\ C_s^{0.579}$	0.3–2.3	0.05–1.4	0.759	Lorenz et al. (1988)

contrast, Brylinsky (1980) found that annual phytoplankton production scaled as mean chlorophyll *a* concentration raised to the power 1.36 in lakes. Whether this difference is due to the wide variation in the temporal scales over which data were averaged (one day for the oceanic data versus a growing season for the lake data), or because of a bias introduced by other uncontrolled variables, or to a fundamental difference in the relation between π and C_s in freshwater and marine systems remains to be evaluated.

Variations in primary production that cannot be accounted for by differences in surface biomass are presumably due to differences in other factors such as temperature, irradiance, nutrient availability, optical properties of the waterbody, vertical distribution and species composition of the phytoplankton (Eppley et al., 1985). Fee (1979) found that annual values of π/C_s for lakes in North-Western Ontario varied with the dominant phytoplankton taxa. The ratio of π/C_s was greatest in lakes dominated by members of the Chlorophyceae, intermediate in lakes dominated by members of the Chrysophyceae, and lowest in lakes dominated by cyanobacteria. Two environmental factors that are amenable to remote sensing and would be expected to influence the relationship between primary production and chlorophyll *a* concentration are irradiance and temperature. To examine the effects of these and other variables on the rate of phytoplankton photosynthesis, it is desirable to define an areal productivity index (π_{chl}), as the ratio of integrated euphotic zone primary production (π) to integrated euphotic zone chlorophyll *a* content (C_i):

$$\pi^{chl} = \pi/C_i \tag{8.6}$$

where $\pi = \int_0^D P(z)\ dz$ and $C_i = \int_0^D C(z)\ dz$, D is the depth of the euphotic zone, $P(z)$ is the rate of primary production at depth z, and $C(z)$ is the

Table 8.4.   Variations in the slope (ψ) of the relationship between the areal productivity index (π_{chl}) and incident irradiance (E) abstracted from the summaries of Platt (1987) and Platt et al. (1988) with additional sources as indicated.

Location	ψ, g $C \cdot E^{-1} \cdot m^2 \cdot g^{-1}$chla	Correlation coefficient
New York Bight	0.43	0.86
Subtropical N. Atlantic	0.44	—
Baffin Bay	0.33	—
Jones Sound, Canadian Arctic	0.36	—
Bedford Basin, Canada	0.45	0.75
English Channel	0.43	—
Southern Ocean	0.37	0.29
Long Island Sound	0.38	0.49
Petpeswick Inlet, Nova Scotia	0.43	0.84
Grand Banks of Newfoundland	0.35	0.92
Midsubartic Pacific	0.66	—
Southeastern United States Shelf (Yoder et al., 1985)	1.5	0.93
Chesapeake and Delaware Bays (Harding et al., 1986)	0.47	—
Northeastern United States (Falkowski et al., 1988)	0.52	—

chlorophyll a concentration at depth z. A linear relation between the areal productivity index (π^{chl}) and incident irradiance (E_o) has been observed by several investigators (Table 8.4). Platt (1986) suggested that the slope of this relation, designated (Ψ), was related to the chlorophyll a-specific absorption coefficient a*, the realized photon yield of photosynthesis in situ (η), the euphotic zone depth (D), and the vertical attenuation coefficient (K) for scalar irradiance as follows:

$$\pi = \pi^{chl} C_i = C_i \Psi E_0 = C_i a^* \eta E_o D K \tag{8.7}$$

From a review of published data, Platt (1986) and Platt et al (1988) concluded that Ψ varied approximately by a factor of two and, based on theoretical grounds, argued for the relative stability of Ψ. In contrast to Platt's (1986) conclusions, Campbell and O'Reilly (1988) found that Ψ varied from 0.1 to 10, with a mean value approximately three times higher than obtained from Platt's (1986) summary. Observations of Yoder et al. (1985) for the southeastern United States continental shelf also yielded a value for Ψ, which was about three times greater than the typical values given by Platt (1986) suggesting some regional variations.

Variability in Ψ may arise from a number of sources, including (1) varia-

Table 8.5.   Dependence of integral primary productivity (π) on chlorophyll concentration (C_s), incident irradiance (E_o), and vertical attenuation coefficient (K).

Location	Slope	Correlation Coefficient
San Francisco Bay		
North Bay (1980)	0.15 ± 0.026	0.72
South Bay (1980)	0.19 ± 0.026	0.88
South Bay (1982)	0.22 ± 0.034	0.78
Hudson River Plume	0.25 ± 0.021	0.94
Puget Sound		
1966	0.16 ± 0.026	0.60
1967	0.15 ± 0.026	0.79
Narragansett Bay and MERL Microcosms		
(Keller, 1988b)	0.15 ± 0.004	0.82

Data modified from a tabular summary by Cole and Cloern (1987) and other sources as indicated. Productivity (π) was regressed on the composite parameter $C_s E_o / K$. The slope of this regression has units mol^{-1} photon.

tions in a^*, (2) variations in η due to variations in the vertical distribution of phytoplankton abundance and in the relationship between phytoplankton biomass and vertical light attenuation, and (3) variations in η that arise from variations in the parameters of the PI curve P_m^{chl} and α^{chl}. It is necessary to consider the full equation for determining π from the depth-dependence of photosynthesis to examine the sources of variability in Ψ. Recent theoretical treatments of this problem which explicitly consider the angular and spectral distributions of the underwater light fields, together with the vertical distribution of phytoplankton abundance (Sathyendranath and Platt, 1989), are beyond the scope of this review.

One variable that is particularly important in determining Ψ is the attenuation coefficient (K). Noting that the euphotic zone depth (D) and attenuation coefficient (K) are related by the equality $DK = \log_e (0.01) = 4.61$ (i.e., $K = D/4.61$) allows Eq. 8.7 to be rewritten as

$$\Psi = a^* \eta 4.61 / K^2 \qquad (8.8)$$

Because of the K^2 dependence of Ψ, explicit consideration of the vertical attenuation coefficient may improve models of integral primary production (π). Explicit consideration of the attenuation coefficient may be particularly important in estuaries where K can vary greatly because of changes in phytoplankton abundance, dissolved organic matter, and suspended sediments. In fact, Cole and Cloern (1987) showed that π scaled with $C_c E_o / K$ in five United States estuaries (Table 8.5). A linear dependence of π on $C_s s E_o / K$ has also been demonstrated in Delaware Bay (Pennock and Sharp, 1986) and Narragansett Bay (Keller, 1988b). Significantly, Pennock and Sharp (1986) found that the slope was three times greater in the summer than during the nonsummer months. They attributed this difference to a change in species composition of the phytoplankton from predominantly microflagellates in summer to diatoms for the rest of the year.

Mechanistic Descriptions of the Relationship between Primary Production and Chlorophyll *a*

For remote sensing applications, it is desirable to relate π to the near-surface chlorophyll *a* concentration (C_s), which is easily measurable, rather than to the integral chlorophyll *a* concentration (C_i), which is not. Strictly, the chlorophyll *a* concentration seen by an airborne or satellite sensor corresponds to a weighted average value of chlorophyll *a* within the first attenuation depth (equal to $1/4.6 = 0.22$ times the euphotic zone depth) and designated C_k. For most practical applications, it is reasonable to assume that $C_s = C_k$. Note that under the assumption of a uniform vertical chlorophyll *a* distribution $C_k = C_s = C_i/D = C_i/4.6K$.

Talling (1957a,b) provided the original formulation for the dependence of π on chlorophyll *a* concentration, incident irradiance, vertical attenuation of scalar irradiance (K) and the PI curve parameters P_m^{chl} and I_k:

$$\pi = C_s P_m^{chl} \log_e (2E_o/I_k)/K \tag{8.9}$$

This formulation has been verified in a range of freshwater environments (Talling, 1965; Bindloss, 1974; Ganf, 1975, Jewson, 1976; Megard et al., 1979; Harris et al., 1980; Oliver and Ganf, 1988), and a similar equation has been developed by Bannister (1974). Note the similarity of Talling's (1957a,b) model (Eq. 8.8) to a recent treatment of the same problem by Platt (1986):

$$\pi = C_s P_m^{chl} \log_e [E/I_k + \sqrt{1 + (E/I_k)^2}] / K \tag{8.10}$$

The difference between the equations of Talling (1957a,b) and Platt (1986) arise from different assumptions about the shape of the PI curve. It can be easily demonstrated that Platt's (1986) model and Talling's (1957a,b) model converge for large values of E_o/I_k.

The depth of the euphotic zone is not explicit in equations relating π

to C_s and the surface irradiance does not enter as a linear multiplicative term in Eqs. 8.9 and 8.10, but rather appears within the logarithmic term at the extreme right-hand side of these equations. How can the different formulations of the dependence of π on E_o given in Eqs. 8.9 and 8.10 be reconciled with the linear dependence implied in Equation 8.7? Noting that $P_m^{chl} = \alpha^{chl} I_k = \phi_m a^* I_k$, and multiplying Eq. 8.9 by E/E (= unity) leads to

$$\pi = [C_s a^*/K] \, [\phi_m I_k \log_e (2E_o/I_k] \, E_o \qquad (8.11)$$

The terms within the first set of square brackets give the proportion of incident irradiance absorbed by phytoplankton, whereas the terms within the second set of square brackets give a measure of the efficiency with which absorbed light is used for photosynthetic carbon fixation, which is equivalent to η in Eq. 8.7. The last term on the right-hand side of Eq. 8.10 is the incident irradiance (E_o). Thus, a logarithmic dependence on E_o is implicit in Eq. 8.7. Several recent theoretical models treat phytoplankton growth rate as a function of irradiance, photon yield, and the absorption coefficient (Bannister, 1979; Shuter, 1979; Kiefer and Mitchell, 1983; Laws et al., 1985). Models of primary production that are conceptually identical to Eq. 8.11 can be derived from these models of phytoplankton growth.

Estimates of primary production from chlorophyll a and incident irradiance are dependent on a number of environmental and physiological factors. The environmental factors involve the partitioning of absorbed photons between phytoplankton and other substances and can be parameterized as the ratio of light absorbed by phytoplankton to total light attenuation. This has two components, the total pigment content within the photic zone and the vertical distribution of pigments. The physiological factors involved include genotypic and phenotypic modifications of the photosynthesis–light response curve (Chapter 7) and can be parameterized by P_m^{chl}, $\alpha^{chl} = a^* \phi_m$, and $I_k = P_m^{chl}/\alpha^{chl}$. The ability to estimate primary production from chlorophyll a will ultimately be limited by the variability of these parameters.

References

Aaronson, S. (1978) Excretion of organic matter by phytoplankton *in vitro*. Limnol. Oceanogr. **23**:838.

Ackleson, S., Balch, W.M., and Holligan, P.M. (1988) White waters of the gulf of Maine. Oceanography **1**:18–22.

Alberte, R.S., Wood, A.M., Kursar, T.A., and Guillard, R.R.L. (1984) Novel phycoerythrins in marine *Synechococcus* spp.: characterization and evolutionary and ecological significance. Plant Physiol. **75**:732–739.

Alpine, A.E., and Cloern, J.E. (1985) Differences in *in vivo* fluorescence yield between three phytoplankton size classes. J. Plankton Res. **7**:381–390.

American Public Health Association (1985) Standard Methods for the Examination of Water and Wastewater, A.E. Greenberg, R.R. Trussel, L.S. Clesceri, and M.A.H. Franson (Eds.), American Public Health Association, Washington DC, pp. 412–426.

Anderson, D.M., and Morel, F.M. (1978) Copper sensitivity of *Gonyaulax tamarensis*. Limnol. Oceanogr. **23**:283–295.

Appleby, G., Colbeck, J., Holdsworth, E.S. and Wadman, H. (1980) β-carboxylation enzymes in marine phytoplankton and isolation and purification of pyruvate carboxylase from *Amphidinium carterae* (Diniphyceae). J. Phycol. **16**:290–295.

Arnold, K.E. and Littler, M.M. (1985) The carbon-14 method for measuring primary productivity. In Handbook of Phycological Methods, Vol. 4, Ecological Field Methods: Macroalgae (M.M. Littler and D.S. Littler) Cambridge University Press, Cambridge, pp. 377–396.

Arnold, K.E., and Manley, S.L. (1985) Carbon allocation in *Macrocystis pyrifera* (Phaeophyta): intrinsic variability in photosynthesis and respiration. J. Phycol. **21**:154–167.

Arnon, D.I., and Barber, J. (1990) Photoreduction of $NADP^+$ by isolated reaction centres of photosystem II: requirement for plastocyanin. Proc. Natl. Acad. Sci. **87**:5–15.

Arnon, D.I., and Tang, M.-S. (1989) Photoreduction of $NADP^+$ by a chloroplast photosystem II preparation: effect of light intensity. Federation of European Biochemical Societies Letters, 253:253–256.

Atkinson, M.J., and Smith, S.V. (1983) C:N:P rations of benthic marine plants, Limnol. Oceanogr. 28:568–574.

Aughey, W.H., and Baum, F.J. (1954) Angular-dependent light scattering—a high resolution recording instrument for the angular range 0.05–140°. J. Opt. Soc. Am. 44:833–837.

Axelsson, L. (1988) Changes in pH as a measure of photosynthesis by marine macroalgae. Mar. Biol. 97:287–294.

Axelsson, L. and Uusitalo, J. (1988) Carbon acquisition strategies for marine macroalgae. I. Utilization of proton exchanges visualized during photosynthesis in a closed system. Mar. Biol. 97:295–300.

Badger, M.R. (1985) Photosynthetic oxygen exchange. Ann. Rev. Plant Physiol. 36:27–53.

Balch, W.M., Holligan, P.M., and Ackleson, S.G. (1988) Optical characteristics of the 1988 coccolithophore bloom in the Gulf of Maine. EOS 69:1132.

Balch, W.M., Eppley, R.W., Abbott, M.R., and Reid, F.M.H. (1989) Bias in satellite-derived measurements due to coccolithophorids and dinoflagellates. J. Plankton Res. 11:575–581.

Bannister, T.T. (1974) A general theory of steady state phytoplankton growth in a nutrient saturated mixed layer. Limnol. Oceanogr., 19:13–30.

Bannister, T.T. (1979) Quantitative description of steady state, nutrient-saturated algal growth, including adaptation. Limnol. Oceanogr. 24:79–96.

Bannister, T.T. (1988) Estimation of absorption coefficients of scattering suspensions using opal glass. Limnol. Oceanogr. 33:607–615.

Banse, K. (1976) Rates of growth respiration and photosynthesis of unicellular algae as related to cell size—a review. J. Phycol. 12: 135–140.

Banse, K. (1982) Cell volumes, maximal growth rates of unicellular algae and ciliates, and the role of ciliates in the marine pelagial. Limnol. Oceanogr. 27:1059–1071.

Barber, J., Malkin, S. and Telfer, A. (1989) The origin of chlorophyll fluorescence *in vivo* and its quenching by the photosystem 2 reaction centre. Phil. Trans. R. Soc. Lond. B323:227–239.

Bassham, J.A., and Calvin, M. (1957) The path of Carbon in Photosynthesis. Prentice-Hall, Englewood Cliffs, NJ.

Bate, G.C., Süeltemeyer, D.F., and Fock, H.P. (1988) $^{16}O_2/^{18}O_2$ analysis of oxygen exchange in *Dunaliella tertiolecta*. Evidence for the inhibition of mitochondrial respiration in the light. Photosyn. Res. 16:219–231

Beardall, J., and Morris, I. (1976) The concept of light intensity adaptation in marine phytoplankton: some experiments with *Phaeodactylum tricornutum*. Mar. Biol. 37:377–387.

Beardall, J., Mukerji, D., Glover, H.E., and Morris, I. (1976) The path of carbon in photosynthesis by marine phytoplankton. J. Phycol. 12:409–417.

Beer, S., and Eshel, A. (1983) Photosynthesis of *Ulva* sp. I. Effects of desiccation when exposed to air. J. Exp. Mar. Biol. Ecol. 70:91–97.

Beer, S., and Levy, I. (1983) Effects of photon fluence rate and light spectrum composition on growth, photosynthesis and pigment relations in *Gracillaria* sp. J. Phycol. **19**:516–522.

Beer, S., Steward, A.J. & Wetzel, R.G. (1982) Measuring chlorophyll *a* and ^{14}C-labelled photosynthate in aquatic angiosperms by use of a tissue solubilizer. Plant Physiol. **69**:54–57.

Bell, L.N. (1985) Energetics of the Photosynthesizing Plant Cell, Harwood Academic Publishers, London, 402 pp.

Bell, D.H., and Hipkins, M.F. (1985) Analysis of fluorescence induction curves from pea chloroplasts. Photosystem 2 reaction centre heterogeneity. Biochim. Biophys. Acta **807**:255–262.

Bender, M., Grande, K., Johnson, K., Marra, J., Williams, P.J.LeB., Sieburth, J., Pilson, M., Langdon, C., Hitchcock, G., Orchardo, J., Hunt, C., Donaghay, P., and Heinemann, K. (1987) A comparison of four methods for determining planktonic community production. Limnol. Oceanogr. **32**:1085–1098.

Bennett, A., and Bogorad, L. (1973) Complementary chromatic adaptation in a filamentous blue-green alga. J. Cell. Biol. **58**:419–435.

Benson, B.B., and Krause, D. Jr. (1984) The concentration and isotopic fractionation of oxygen dissolved in freshwater and seawater in equilibrium with the atmosphere. Limnol. Oceanogr. **29**:620–632.

Benson, B.B., and Parker, P.D.M. (1961) Relations among the solubilities of nitrogen, argon and oxygen in distilled water and sea water. J. Phys. Chem. **65**:1489–1496.

Berger, M., and Bate, G.C. (1986) Measurement of photosynthetic rate in marine macrophytes, a comparison of carbon based methods. Bot. Mar. **29**:177–183.

Berner, T., Dubinsky, Z., Wyman, K., and Falkowski, P.G. (1989) Photoadaptation and the "package" effect in *Dunaliella tertiolecta* (Chlorophyceae). J. Phycol. **25**:70–78.

Berry, J.A. (1988) Studies of mechanisms affecting the fractionation of carbon isotopes in photosynthesis. *In* Stable Isotopes in Ecological Research, P.W. Rundel, J.R. Ehleringer, and K.A. Nagy (Eds.), Springer-Verlag, New York, pp. 82–94.

Beyers, R.J. (1963) The metabolism of twelve aquatic laboratory microecosystems. Écol. Monogr. **33** (4):282–306.

Beyers, R.J., and H.T. Odum (1959) The use of carbon dioxide to construct pH curves for the measurement of productivity. Limnol. Oceanogr. **4**:499–502.

Bickford, E.D., and Dunn, S. (1972) Lighting for plant growth. Kent State University Press, Kent, OH.

Bidigare, R.R., Morrow, J.H., and Kiefer, D.A. (1989) Derivative analysis of spectral absorption by photosynthetic pigments in the western Sargasso Sea. J. Mar. Res. **47**:323–341.

Bidigare, R.R., Smith, R.C., Baker, K.S. and Marra, J. (1987) Oceanic primary production estimates from measurements of spectral irradiance and pigment concentrations. Global Biogeochemical Cycles, **1**:171–186.

Bidwell, R.G.S. (1977) Photosynthesis and light and dark respiration in freshwater algae. Can. J. Bot. **55**:809–819.

Bidwell, R.G.S., and McLachlin, J. (1985) Carbon nutrition of seaweeds: photosynthesis, photorespiration and respiration. J. Exp. Mar. Biol. Ecol. 86:15–46.

Bilger, W., and Schreiber, U. (1986) Energy-dependent quenching of dark-level chlorophyll fluorescence in intact leaves. Photosynth. Res. 10:303–308.

Bindloss, M.E. (1974) Primary productivity of phytoplankton in Loch Leven, Kinross. Proc. R. Soc. Edinburgh B74:157–181.

Birkebak, R.C., and Cho, S.H. (1967) Integrating sphere blockage effects. American Society of Mechanical Engineers, Paper #67.

Björkman, O. (1987) Low-temperature chlorophyll fluorescence in leaves and its relationship to the photon yield of photosynthesis in photoinhibition. In: Photoinhibition, D.J. Kyle, C.B. Osmond, and C.J. Arntzen (Eds.), Elsevier, Amsterdam, pp. 123–145.

Björkman, O., and Demmig, B. (1987) Photon yield of O_2 evolution and chlorophyll fluorescence characteristics at 77 K among vascular plants of diverse origins. Planta 170:489–504.

Bjornsen, P.K. (1988) Phytoplankton exudation of organic matter: Why do healthy cells do it? Limmol. Oceanogr. 33:151–154.

Blackman, F.F. (1905) Optima and limiting factors. Ann. Bot. 19:281–295.

Blasco, D., Packard, T.T., and Garfield, P.C. (1982) Size dependence of growth rate, respiratory electron transport system activity and chemical composition in marine diatoms in the laboratory. J. Phycol. 18:58–63.

Blinks, L.R. and Skow, R.K. (1938) The time course of photosynthesis as shown by a rapid electrode method for oxygen. Proc. Natl. Acad. Sci. 24:420–427.

Bolhar-Nordenkampf, H.R., Long, S.P., Baker, N.R., Oquist, G., Schreiber, U., and Lechner, E.G. (1989) Chlorophyll fluorescence as a probe of the photosynthetic competence of leaves in the field: a review of current instrumentation. Funct. Ecol. 3:497–514.

Borowitzka, M.A. (1977) Algal calcification. Oceanogr. Mar. Biol. Annu. Rev. 15:189–223.

Borowitzka, M.A., and Larkum, A.W.D. (1976) Calcification in the green alga Halimeda. II. The exchange of Ca^{2+} and the occurrence of gradients in calcification and photosynthesis. J. Exp. Bot. 27:864–878.

Boulding, E.G., and Platt, T. (1986) Variation in photosynthetic rates among individual cells of a marine dinoflagellate. Mar. Ecol. Prog. Ser. 29:199–203.

Boussiba, S., and Richmond, A.E. (1980) C-phycocyanin as a storage protein in the blue-green alga Spirulina platensis. Arch. Microbiol. 125:143–147.

Brackett, F.S., Olson, R.A., and Crickard, R.G. (1953a) Respiration and intensity dependence of photosynthesis in Chlorella. J. Gen. Physiol. 36:529–561.

Brackett, F.S., Olson, R.A., and Crickard, R.G. (1953b) Time course and quantum efficiency of photosynthesis in Chlorella. J. Gen. Physiol. 36:563–579.

Bradbury, M., and Baker, N.R. (1981) Analysis of the slow phase of the in vivo chlorophyll a induction curve. Changes in the redox state of photosystem II electron acceptors and fluorescence emission from photosystems I and II. Biochim. Biophys. Acta 635:542–551.

Bradshaw, A.L., and Brewer, P.G. (1988a) High precision measurements of alkalinity and total carbon dioxide in seawater by potentiometric titration. 1. Presence of unknown protolyte(s)? Mar. Chem. 23:69–86.

Bradshaw, A.L., and Brewer, P.G. (1988b) High precision measurements of alkalinity and total carbon dioxide in seawater by potentiometric titration. Mar. Chem. 24:155–162.

Bradshaw, A.L., Brewer, P.G., Shafer, D., and Williams, R.T. (1981) Measurements of total carbon dioxide and alkalinity by potentiometric titration in the GEOSECS program. Earth Planet. Sci. Lett. 55:99–115.

Brand, L.E. (1981) Genetic variability in reproduction rates in marine phytoplankton populations. Evolution 35:1117–1127.

Brand, L.E. (1985) Low genetic variability in reproduction rates in populations of Prorocentrum micans Ehrenb. (Dinophyceae) over Georges Bank. J. Exp. Mar. Biol. Ecol. 88:55–65.

Brand, L.E. and Guillard, R.R.L. (1981) The effects of continuous light and light intensity on the reproduction rates of twenty-two species of marine phytoplankton. J. Exp. Mar. Biol. Ecol. 50:119–132.

Brand, L.E., Guillard, R.R.L. and Murphy, L.S. (1981) A method for the rapid and precise determination of acclimated phytoplankton reproduction rates. J. Plankton Res. 3:193–201.

Brand, L.E., Sunda, W.G., and Guillard, R.R.L. (1986) Reduction of marine phytoplankton reproductive rates by copper and cadmium. J. Exp. Mar. Biol. Ecol. 96:225–250.

Brechignac, F., and Furbank, R.T. (1987) On the nature of the oxygen uptake in the light by Chondrus crispus. Effects of inhibitors, temperature and light intensity. Photosynth. Res. 11:45–59.

Brechignac, F., Ranger, C., Andre, M., Daguenet, A., and Massimino, D. (1987) Oxygen exchanges in marine macroalgae. In Progress in Photosynthesis Research, Vol III, J. Biggens (Ed.), Martinus Nijhoff Publishers, Dordrecht, pp. 657–660.

Briantis, J.M., Vernotte, C., Krause, G.H., and Weis, E. (1986). Chlorophyll a fluorescence of higher plants: chloroplasts and leaves. In Light Emission by Plants and Bacteria, Govindjee, J. Amesz, and D.C. Fork (Eds.), Academic Press, Orlando, pp. 539–583.

Bricaud, A., and Morel, A. (1986) Light attenuation and scattering by phytoplanktonic cells: a theoretical modeling. Appl. Opt. 25:571–580.

Bricaud, A. and Morel, A. (1987) Atmospheric corrections and interpretation of marine radiances in CZCS imagery: use of a reflectance model. Oceanol. Acta. 7:33–50.

Bricaud, A., Bedhomme, A.L., and Morel, A. (1988) Optical properties of diverse phytoplankton species: experimental results and theoretical interpretation. J. Plankton Res. 10:851–873.

Bricaud, A., Morel, A., and Prieur, L. (1983) Optical efficiency factors of some phytoplankters. Limnol. Oceanogr. 28:816–832.

Brinkhaus, B.H., Tempel, N.R., and Jones, (1976) Photosynthesis and respiration of exposed salt marsh fuccoids. Mar. Biol. 34:349–359.

Britz, S.J., and Briggs, W.R. (1987) Chloroplast movement and light transmission in *Ulva*: The sieve effect in a light-scattering system. Acta Physiol. Plant **9**:149–162.

Broeker, W.S., and Peng, T.-H. (1982) Tracers in the sea, The Lamont-Doherty Geological Laboratory, Palisades, New York.

Broenkow, W.W., and Cline, J.D. (1969) Colorimetric determination of dissolved oxygen at low concentrations. Limnol. Oceanogr. **14**:450–454.

Brown, A.H. (1953) The effects of light on respiration using isotopically enriched oxygen. Am. J. Bot. **40**:719–729.

Brown, A.H., and Good, N. (1955) Photochemical reduction of oxygen in chloroplast preparations and green plant cells. I. The study of oxygen exchanges *in vitro* and *in vivo*. Arch. Biochem. Biophys. **57**:340–354.

Brown, A.H., and Webster, G.C. (1953) The influence of light on the rate of respiration of the blue-green alga *Anabaena*. Am. J. Bot. **40**:753–759.

Brown, R.G.W. (1987) Dynamic light scattering using monomode optical fibres. Appl. Opt. **26**:4846–4851.

Browse, J.A. (1979) An open-circuit infrared gas analysis system for measuring aquatic plant photosynthesis at physiological pH. Aust. J. Plant Physiol. **6**:493–498.

Browse, J.A. (1985) Measuring photosynthesis by infrared gas analysis. *In* Handbook of Phycological Methods, Vol. 4, Ecological Field Methods: Macroalgae, M.M. Littler and D.S. Littler (Eds.), Cambridge University Press, Cambridge, pp. 397–414.

Brujewicz, S.W. (1936) Estimation of the Organic Matter Production in the Sea (Caspian Sea). Academy Science USSR, Moscow, pp. 281–300 (in Russian, English summaries).

Brylinsky, M. (1977) Release of dissolved organic matter by some marine macrophytes. Mar. Biol. **27**:137–141.

Brylinsky, M. (1980) Estimating the productivity of lakes and reservoirs. *In* The Functioning of Freshwater Ecosystems, E.D. LeCren and R.H. Lowe-McConnell, (Eds), Cambridge University Press, New York, pp. 411–453.

Buckingham, S., Walters, C.J., and Kleiber, P. (1975) A procedure for estimating gross production, net production and algal carbon content using ^{14}C. Int. Ver. Theor. Angew. Limnol. Verh. **19**:32–38.

Buesa, R.J. (1980) Photosynthetic quotient of marine plants. Photosynthetica **14**:337–342.

Burger-Wiersma, T. and Post, A.F. (1989) Functional analysis of the photosynthetic apparatus of *Proclorothrix hollandica* (Prochlorales), a chlorophyll *b* containing procaryote. Plant Physiol., **91**:770–774.

Burkhill, P.H., Mantoura, R.F.C., Lewellyn, C.A., and Owens, N.J.P. (1987) Microzooplankton grazing and selectivity of phytoplankton in coastal waters. Mar. Biol. **93**:581–590.

Burns, B.D., and Beardall, J. (1987) Utilization of inorganic carbon by marine microalgae. J. Exp. Mar. Biol. Ecol. **107**:75–86.

Burris, J.E. (1981) Effects of oxygen and inorganic carbon concentrations on the photosynthetic quotients of marine algae. Mar. Biol. **65**:215–219.

Butler, W.L. (1964) Absorption spectroscopy *in vivo*: theory and application. Annu. Rev. Plant Physiol. **15**:451–470.

Butler, W.L. (1977) Chlorophyll fluorescence: a probe of electron transfer and energy transfer. *In* Photosynthesis I. Photosynthetic Electron Transport and Photophosphorylation, Encyclopedia of Plant Physiology, A. Trebst and M. Avron (Eds.), Springer-Verlag, Berlin, pp. 149–167.

Butler, W.L. (1978) Energy distribution in the photochemical apparatus of photosynthesis. Annu. Rev. Plant Physiol. **29**:345–378.

Butler, W.L. (1979) Fourth derivative spectra. Meth. in Enzymol. **16**:501–515.

Caemmerer, S., and Farquhar, G. (1981) Some relationships between the biochemistry of photosynthesis and the gas-exchange of leaves. Planta **153**:376–387.

Campbell, J.W., and O'Reilly, J.E. (1988) Role of satellites in estimating primary productivity on the northwest Atlantic continental shelf. Cont. Shelf. Res. **8**:179–204.

Carder, K.L., and Steward, R.G. (1985) A remote-sensing reflectance model of a red-tide dinoflagellate bloom off west Florida. Limnol. Oceanog. **30**:286–298.

Carlson, D.J., and Carlson, M.L. (1984) Reassessment of exudation by fucoid macroalgae. Limnol. Oceanogr. **29**:1077–1087.

Caron, D.A. (1983) Technique for enumeration of heterotrophic and phototrophic nanoplankton using epifluorescence microscopy and comparison with other procedures. Appl. Environ. Microb. **46**:491–498.

Carpenter, J.H. (1965a) The accuracy of the Winkler method for dissolved oxygen analysis. Limnol. Oceanogr. **10**:135–140.

Carpenter, J.H. (1965b) The Chesapeake Bay Institute Technique for the Winkler dissolved oxygen method. Limnol. Oceanogr. **10**:141–143.

Carpenter, E.J., and Lively, J.S. (1980) Review of estimates of algal growth using ^{14}C tracer techniques. *In* Primary Productivity in the Sea, P.G. Falkowski (Ed.), Plenum Press, New York, pp. 161–178.

Chalker, B.E. (1980) Modeling light saturation curves for photoysnthesis: an exponential function. J. Theor. Biol. **84**:205–215.

Chalker, B.E. (1981) Simulating light-saturation curves for photosynthesis and calcification by reef-building corals. Mar. Biol. **63**:503–507.

Chan, A.T. (1978) Comparative physiological study of marine diatoms and dinoflagellates in relation to irradiance and cell size. I. Growth under continuous light. J. Phycol. **16**:396–402.

Chan, A.T. (1980) Comparative physiological study of marine diatoms and dinoflagellates in relation to irradiance and cell size. II. Relationship between photosynthesis, growth, and carbon/chlorophyll *a* ratio. *J. Phycol.* **16**:428–432.

Chandler, M.T., and Vidaver, W.E. (1971) Stationary platinum electrode for measurement of O_2 exchange by biological systems under hydrostatic pressure. Rev. Sci. Instr. **42**:143–146.

Chapman, A.R.O. and Lindley, J.E. (1980) Seasonal growth of *Laminaria solidungula* in the Canadian high arctic in relation to irradiance and dissolved nutrient concentrations. Mar. Biol. **57**:1–5.

Chrost, R.J., and Faust, M.A. (1983) Organic carbon release by phytoplankton:

its composition and utilization by bacterioplankton. J. Plankton Res. **5**:477–493.

Cleveland, J.S., and Perry, M.J. (1987) Quantum yield, relative specific absorption and fluorescence in nitrogen-limited *Chaetocerous gracilis*. Mar. Biol. **94**:489–497.

Cohen-Bazire, G., and Bryant, D.A. (1981) Phycobilisomes: composition and structure. *In* The Biology of Cyanobacteria. N.C. Carr and B.A. Whitton (Eds.) Blockwell Scientific Publishers, Oxford, pp. 143–190.

Cole, B.E., and Cloern, J.E. (1987) An empirical model for estimating phytoplankton productivity in estuaries. Mar. Ecol. Prog. Ser. **36**:299–305.

Cole, J.J., McDowell, W.H., and Likens, G.E. (1984) Sources and molecular weight of 'dissolved' organic carbon in an oligotrophic lake. Oikos **42**:1–9.

Conrad, R., and Seiler, W. (1980) Photooxidative production and microbial consumption of carbon monoxide in seawater. FEMS Microb. Lett. **9**:61–64.

Côté, B., and Platt, T. (1983) Day-to-day variations in the spring-summer photosynthetic parameters of coastal marine phytoplankton. Limnol. Oceanogr. **28**:320–344.

Côté, B., and Platt, T. (1984) Utility of the light-saturation curve as an operational model for quantifying the effects of environmental conditions on phytoplankton photosynthesis. Mar. Ecol. Prog. Ser. **18**:57–66.

Coutinho, R., and Zingmark, R. (1987) Diurnal photosynthetic responses to light by macroalgae. J. Phycol. **23**:336–343.

Craigie, J.S., McLachlan, J., and Majak, W. (1966) Photosynthesis in algae. II. Green algae with special reference to *Dunaliella* spp. and *Tetraselmis* spp. Can. J. Bot. **44**:1247–1254.

Cramer, M., and Myers, J. (1948) Nitrate reduction and assimilation in *Chlorella*. J. Gen. Physiol. **32**:93–102.

Cuhel, R.L., Ortner, P.B., and Lean, D.R.S. (1984) Night synthesis of protein by algae. Limnol. Oceanogr. **29**:745–762.

Culberson, C.H., and Huang, S. (1987) Automated amperometric oxygen titration. Deep-Sea Res. **34**:875–880.

Cullen, J.J. (1982) The deep chlorophyll maximum: comparing profiles of chlorophyll *a*. Can. J. Fish. Aquat. Sci. **39**:791–803.

Cullen, J.J. and Lewis, M.R. (1988) The kinetics of algal photoadaptation in the context of vertical mixing. J. Plankton Res. **10**:1039–1063.

Curtis, P.J., and Megard, R.O. (1987) Interactions among irradiance, oxygen evolution and nitrite uptake by *Chlamydomonas* (Chlorophyceae). J. Phycol. **23**:608–613.

Das, M., Rabinowitch, E., Szalay, L., and Papageorgiou, G. (1967) The "sieve effect" in *Chlorella* suspensions. J. Phys. Chem. **71**:3543–3549.

Davenport, J.B., and Maguire, B. (1984) Quantitative grain density autoradiography and the intraspecific distribution of primary productivity in phytoplankton. Limnol. Oceanogr. **29**:410–416.

Davison, I.R. (1987) Adaptation of photosynthesis in *Laminaria saccharina* (Phaeophyta) to changes in growth temperature. J. Phycol. **23**:273–283.

Davies, B.H. (1976) Carotenoids. *In* Chemistry and Biochemistry of Plant Pigments, Vol. 2, Goodwin T. (Ed.), Academic Press, London, pp. 38–165.

Dawes, C.J. (1985) Respirometry and manometry. pp. 329–348 In Handbook of Phycological Methods. Ecological Field Methods: Macroalgae, M.M. Littler and D.S. Littler (Eds.), Cambridge University Press, Cambridge, pp. 329–348.

Delieu, T., and Walker, D.A. (1981) Polarographic measurement of photosynthetic O_2 evolution by leaf discs. New Phytologist 89:165–175.

De Loura, I.C., Dubacq, J.P., and Thomas, J.C. (1987) The effects of nitrogen deficiency on pigments and lipids of cyanobacteria. Plant Physiol. 83:838–843.

Demmig, B., Winter, K., Kniger, A., and Czygan, F.-C. (1987) Photoinhibition and zeaxanthin formation in intact leaves. A possible role of the xanthophyll cycle in the dissipation of excess light energy. Plant Physiol. 84:218–224.

Descolas-Gros, C., and Fontugne, M.R. (1985) Carbon fixation in marine phytoplankton: carboxylase activities and stable carbon-isotope ratios; physiological and paleoclimatological aspects. Mar. Biol. 87:1–6.

Descolas-Gros, C., and Fontugne, M. (1990) Stable carbon isotope fractionation by marine phytoplankton during photosynthesis. Plant Cell Environ. 13:207–218.

DiTullio, G., and Laws, E.A. (1984) Estimates of phytoplankton N uptake based on $^{14}CO_2$ incorporation into protein. Limnol. Oceanogr. 28:177–185.

DiTullio, G., and Laws, E.A. (1986) Diel periodicity of nitrogen and carbon assimilation in five species of marine phytoplankton: accuracy of methodology for predicting N-assimilation rates and N/C composition ratios. Mar. Ecol. Prog. Ser. 32:123–132.

Dodd, W.A., and Bidwell, R.G.S. (1971) Photosynthesis and gas exchange of *Acetaburlaria* chloroplasts in an artificial leaf. Nature, 234:45–47.

Dor, I., and Levy, I. (1987) Single-sample technique for measuring oxygen production and consumption in macrophytic algae. Aquat. Bot. 27:323–331.

Doty, M.S., and Oguri, M. (1957) Evidence for a photosynthetic daily periodicity. Limnol. Oceanogr. 2:37–40.

Doucha, J., and Kubin, S. (1976) Measurement of *in vivo* absorption spectra of microscopic algae using bleached cells as a reference sample. Arch. Hydrobiol. Suppl. Bd. 49:199–213.

Douglas, D.J. (1984) Microautoradiography-based enumeration of photosynthetic picoplankton with estimates of carbon-specific growth rates. Mar. Ecol. Prog. Ser. 14:223–228.

Dring, M.J. (1981) Chromatic adaptation of photosynthesis in benthic marine algae: an examination of its ecological significance using a theoretical model. Limnol. Oceanogr. 26:271–284.

Dring, M.J., and Jewson, D.H. (1982) What does ^{14}C uptake by phytoplankton really measure? A theoretical modelling approach. Proc. R. Soc. Lond. B214:351–368.

Dring, M.J., and Lüning, K. (1985) Emerson enhancement effect and quantum yield of photosynthesis for marine macroalgae in simulated underwater light fields. Mar. Biol. 87:109–117.

Dromgoole, F.I. (1987) Photosynthesis of marine algae in fluctuating light. I.

Adjustment of rate in constant and fluctuating light regimes. Funct. Ecol. 1:377–386.

Droop, M.R. (1966) Vitamin B_{12} and marine ecology. III. An experiment with a chemostat. J. Mar. Biol. Assoc. UK **46**:659–671.

Droop, M.R., Mickelson, M.J., Scott, J.M., and Turner, M.F (1982) Light and nutrient status of algal cells. J. Mar. Biol. Assoc. UK **62**:403–434.

Dubinsky, Z., Falkowski, P.G., and Wyman, K. (1986) Light harvesting and utilization by phytoplankton. Plant Cell Physiol. **27**:1335–1349.

Dubinsky, Z., Falkowski, P.G., Post, A.F., and van Hes, U.M. (1987) A system for measuring phytoplankton photosynthesis in a defined light field with an oxygen electrode. J. Plankton Res. **9**:607–612.

Duke, C.S., Litaker, R.W., and Ramus, J. (1987) Seasonal variation in RuPCase activity and N allocation in the chlorophyte seaweeds *Ulva curvata* (Kutz.) DeToni and *Codium decorticatum* (Woodw.) Howe. J. Exp. Mar. Biol. Ecol. **112**:145–164.

Duncan, M.J., and Harrison, P.J. (1982) Comparison of solvents for extracting chlorophylls from marine macrophytes. Bot. Mar. **25**:445–447.

Dunstan, W.M. (1973) A comparison of the photosynthesis-light intensity relationship in phylogenetically different marine microalgae. J. Exp. Mar. Biol. Ecol. **13**:181–187.

Dunton, K.H. (1985) Growth of dark-exposed *Laminaria saccharina* (L.) Lamour, and *Laminaria solidungula* J.Ag. (Laminariales, Phaeophyta) in the Alaskan Beaufort Sea. J. Exp. Mar. Biol. Ecol. **94**:181–189.

Duysens, L.N.M. (1956) The flattening of the absorption spectrum of suspensions, as compared to that of solutions. Biochim. Biophys. Acta **19**:1–12.

Duysens, L.N.M. (1986) Introduction to (Bacteria) chlorophyll emission: a historical perspective. *In* Light Emission by Plants and Bacteria, Govindjee, J. Amesz and D.C. Fork (Eds.), Academic Press, Orlando, pp. 3–28.

Duysens, L.N.M. (1989) The discovery of the two photosynthetic systems: a personal account. Photosynth. Res. **21**:61–179.

Duysens, L.N.M., and Sweers, H.E. (1963) Mechanisms of the two photochemical reactions in algae as studied by means of fluorescence. *In* Studies on Microalgae and Photosynthetic Bacteria, University of Tokyo Press, Tokyo, pp. 353–372.

Ehleringer, J., and Bjorkman, O. (1977) Quantum yields for CO_2 uptake in C_3 and C_4 plants. Plant Physiol. **59**:86–89.

Emerson, R., and Arnold, W. (1932) The photochemical reaction in photosynthesis. Gen. Physiol. **16**:191–205.

Emerson, R., and Chalmers, R. (1957) On the efficiency of photosynthesis above and below compensation of respiration. *In* Research in Photosynthesis, Interscience, New York, pp. 349–352.

Emerson, R., Chalmers, R., and Cederstrand, C. (1957) Some factors influencing the long-wave limit of photosynthesis. Botany **43**:133–143.

Emerson, R., and Lewis, C.M. (1942) The photosynthetic efficiency of phycocyanin in *Chroococcus* and the problem of carotenoid participation in photosynthesis. J. Gen. Physiol. **25**:579–595.

Emerson, R., and Lewis, C.M. (1943) The dependence of the quantum yield of *Chlorella* photosynthesis on wavelength of light. Am. J. Bot. **30**:165–178.

Eppley, R.W. (1968) An incubation method for estimating the carbon content of phytoplankton in natural samples. Limnol. Oceanogr. **13**:574–582.

Eppley, R.W. (1972) Temperature and phytoplankton growth in the sea. Fisher, Bull. **70**:1063–1085.

Eppley, R.W., Renger, E.H., Venrick, E.L., and Mullin, M.M., (1973) A study of plankton dynamics and nutrient cycling in the central gyre of the North Pacific Ocean. Limnol. Oceanogr. **18**:534–551.

Eppley, R.W., Stewart, E., Abott, M.R., and Heyman, U. (1985). Estimating ocean primary production from satellite chlorophyll: introduction and regional statistics for the Southern California Bight. J. Plankton Res. **7**:57–70.

Evans, J.R. (1987) The dependence of quantum yield on wavelength and growth irradiance. Aust. J. Plant Physiol. **14**:69–79.

Exton, R.J., Houghton, W.M., Esaias, W., Harris, R.C., Farmer, F.H., and White, H.H. (1983) Laboratory analysis of techniques for remote sensing of estuarine parameters using laser excitation. Appl. Opt. **22**:54–64.

Falkowski, P.G., Flagg, C.N., Rowe, G.T., Smith, S.L., Whitledge, T.E. and Wirick, C.G. (1988) The fate of a spring phytoplankton bloom: export or oxidation? Continental Shelf Research 8, 457–484.

Falkowski, P.G., and Kiefer, D.A. (1985) Chlorophyll *a* fluorescence in phytoplankton: relationship to photosynthesis and biomass. J. Plankton Res. **7**:715–731.

Falkowski, P.G., and Owens, T.G. (1980) Light-shade adaptation: two strategies in marine phytoplankton. Plant Physiol. **66**:592–595.

Falkowski, P.G., Owens, T.G., Ley, A.C., and Mauzerall, D.C. (1981) Effects of growth irradiance levels on the ratio of reaction centers in two species of marine phytoplankton. Plant Physiol. **68**:969–973.

Falkowski, P.G., Dubinsky, Z., and Wyman, K. (1985) Growth-irradiance relationships in phytoplankton. Limnol. Oceanogr. **30**:311–321.

Falkowski, P.G., Wyman, K., Ley, A.C., and Mauzerall, D.C. (1986) Relationship of steady-state photosynthesis to fluorescence in eucaryotic algae. Biochim. Biophys. Acta **849**:183–192.

Falkowski, P.G., Kolber, Z., and Fujita, Y. (1988) Effect of redox state on the dynamics of photosystem 2 during steady-state photosynthesis in eucaryotic algae. Biochim. Biophys. Acta **933**:432–443.

Falkowski, P.G., Sukenik, A., and Herzig, R. (1989) Nitrogen limitation in *Isochrysis galbana* (Haptophyceae). II. Relative abundance of chloroplast proteins. J. Phycol. **25**:471–478.

Farquhar, G.D., Ehleringer, J.R., and Hubick, K.T. (1989) Carbon isotope discrimination and photosynthesis. Ann. Rev. Plant Physiol. **40**:503–537.

Fasham, M.J.R., and Platt, T. (1983) Photosynthetic response of phyotplankton to light: a physiological model. Proc. R. Soc. Lond. **B219**:355–370.

Faust, M.A., and Norris, K.H. (1982) Rapid *in vivo* spectrophotometric analysis of chlorophyll pigments in intact phytoplankton cultures. Brit. Phycol. J. **17**:351–361.

Faust, M.A., and Norris, K.H. (1985) *In vivo* spectrophotometric analysis of photosynthetic pigments in natural populations of phytoplankton. Limnol. Oceanogr. **30**:1316–1322.

Fee, E.V. (1979) A relation between lake morphometry and primary productivity and its use in interpreting whole-lake eutrophication experiments. Limnol. Oceanogr. **24**:401–416.

Fenchel, T. (1974) Intrinsic rate of natural increase: the relationship with body size. Oecologia **14**:317–326.

Fenton, G.E., and Ritz, D.A. (1989) Spatial variability of $^{13}C:^{12}C$ and D:H in *Ecklonia radiata* (C.Ag.) J. Agardh (Laminariales). Est. Coast. Shelf Sci. **28**:95–101.

Ferek, R.J., and Andreae, M.O. (1984) Photochemical production of carbonyl sulphide in marine surface waters. Nature **307**:148–150.

Filbin, G.J., and Hough, R.A. (1984) Extraction of ^{14}C-labelled photosynthate from aquatic plants with dimethyl sulfoxide (DMSO). Limnol. Oceanogr. **29**:426–428.

Fitzwater, S.E., Knauer, G.A., and Martin, J.H. (1982) Metal contamination and its effect on primary production measurements. Limnol. Oceanogr. **27**:544–551.

Forbes, J.R., Denman, K.L., and Mackas, D.L. (1986) Determinations of photosynthetic capacity in coastal marine phytoplankton: effects of assay irradiance and variability of photosynthetic parameters. Mar. Ecol. Prog. Ser. **32**:181–191.

Forbes, M.A., and Canny, M.J. (1981) Spectral properties of brown algae. Proc. Int. Seaweed Symp. **8**:188–192.

Fork, D.C. (1963) Observations on the function of chlorophyll *a* and accessory pigments in photosynthesis. *In* Photosynthetic Mechanisms of Green Plants, B. Kok and A.T. Jagendorf (Eds.), National Academy of Sciences, Washington, DC, pp. 352–361.

Fork, D.C., and Mohanty, P. (1986) Fluorescence and other characteristics of blue-green algae (cyanobacteria), red algae and cryptomonads. *In* Light Emission by Plants and Bacteria, Govindjee, J. Amesz, and D.C. Fork (Eds.), Academic Press, Orland, pp. 453–496.

Fork, D.C., and Satoh, K. (1986) The control by state transitions of the distribution of excitation energy in photosynthesis. Annu. Rev. Plant Physiol. **37**:335–361.

Fork, D.C., Oquist, G., and Hoch, G.E. (1982) Fluorescence emission from photosystem 1 at room temperature in the red alga *Porphyra perforata*. Plant Sci. Lett. **14**:249–254.

Foy, R. H. (1980) The influence of surface to volume ratio on the growth rates of planktonic blue-green algae. Brit. Phycol. J. **15**:279–289.

Foy, R.H., and Gibson, C.E. (1982a) Photosynthetic characteristics of planktonic blue-green algae: the response of twenty strains grown under high and low light. Brit. Phycol. J. **17**:169–182.

Foy, R.H., and Gibson, C.E. (1982b) Photosynthetic characteristics of planktonic blue-green algae: changes in photosynthetic capacity and pigmentation of *Oscillatoria redekei* Van Goor under high and low light. Brit. Phycol. J. **17**:183–193.

Fuhrman, J. (1987) Close coupling between release and uptake of dissolved free amino acids in seawater studied by an isotope dilution approach. Mar. Ecol. Prog. Ser. **37**:45–52.

Fukshansky, L. (1978) On the theory of light absorption in nonhomogenous objects. J. Math. Biol. **6**:177–196.

Fukshansky, L. (1981) Optical properties of plants. *In* Plants and the Daylight Spectrum, H. Smith (Ed.), Academic Press, London, pp. 21–40.

Gaarder, T., and Gran, H.H. (1927) Investigations of the production of plankton in the Oslofjord. Rappt. Cons. Explor. Mer. **42**:3–48.

Gallagher, J.C. (1982) Physiological variation and electrophoretic banding patterns of genetically different seasonal populations of *Skeletonema costatum* (Bacillariophceae). J. Phycol. **18**:148–162.

Gallagher, J.C., Wood, A.M., and Alberte, R.S. (1984) Ecotypic differentiation in the marine diatom *Skeletonema costatum*: influence of light intensity on the photosynthetic apparatus. Mar. Biol. **82**:121–134.

Gallegos, C.L., and Platt, T. (1985) Vertical advection of phytoplankton and productivity estimates: a dimensional analysis. Mar. Ecol. Prog. Ser. **26**:125–134.

Gallegos, C.L., Hornberger, G.M., and Kelly, M.G. (1980) Photosynthesis-light relationships of a mixed culture of phytoplankton in fluctuating light. Limnol. Oceanogr. **25**:1082–1092.

Gallegos, C.L., Church, M.R., Kelly, M.G., and Hornberger, G.M. (1983) Asynchrony between rates of oxygen production and inorganic carbon uptake in a mixed culture of phytoplankton. Arch. Hydrobiol. **96**:164–175.

Ganf, G.G. (1975) Photosynthetic production and irradiance-photosynthesis relationships of the phytoplankton from a shallow equatorial lake (Lake George, Uganda). Oecologia **18**:165–183.

Gantt, E. (1981) Phycobilisomes. Annu. Rev. Plant Physiol. **32**:327–347.

Gausman, H.W. (1985) Plant leaf optical properties in visible and near-infrared light. Graduate Studies, Texas Tech University, No. 29.

Gauthier, C., and Katsaros, K.B. (1984) Insolation during STREX: Comparisons between surface measurements and satellite estimates. J. Geophys. Res. **89**:779–788.

Gavis, J., Guillard, R.R.L., and Woodward, B.L. (1981) Cupric ion activity and the growth rate of phytoplankton clones isolated from different marine environments. J. Mar. Res. **39**:315–333.

Geacintov, N.E., and Breton, J. (1987) Energy transfer and fluorescence mechanisms in photosynthetic membranes. CRC Crit. Rev. Plant Sci. **5**:1–44.

Geider, R.J. (1987) Light and temperature dependence of the carbon to chlorophyll *a* ratio in microalgae and cyanobacteria: implications for physiology and growth of phytoplankton. New Phytol. **106**:1–34.

Geider, R.J. (1988) Abundances of autotrophic and heterotrophic nanoplankton and the size distribution of microbial biomass in the southwestern North Sea in October 1986. J. Exp. Mar. Biol. Ecol. **123**:127–145.

Geider, R.J. (1990) Comment: The relationship between steady state phytoplankton growth and photosynthesis. Limnol. Oceanogr. **35**:971–972.

Geider, R.J., and Osborne, B.A. (1987) Light absorption by a marine diatom: experimental observations and theoretical calculations of the package effect in a small *Thalassiosira* species. Mar. Biol. **96**:299–308.

Geider, R.J., and Osborne, B.A. (1989) Respiration and microalgal growth: a review of the quantitative relationship between dark respiration and growth. New Phytologist **112**:327–341.

Geider, R.J., and Platt, T. (1986) A mechanistic model of photoadaptation in microalgae. Mar. Ecol. Prog. Ser. **30**:85–92.

Geider, R.J., Osborne, B.A., and Raven, J.A. (1984) Light dependence of growth and photosynthesis in *Phaeodacctylum tricornutum* (Bacillariophyceae). J. Phycol. **21**:609–619.

Geider, R.J., Osborne, B.A., and Raven, J.A. (1986) Growth, photosynthesis and maintenance metabolic cost in the diatom *Phaeodactylum tricornutum* at very low light levels. J. Phycol. **22**:39–48.

Geider, R.J., T. Platt and J.A. Raven (1986) "Size dependence of growth and photosynthesis in diatoms: a synthesis." Mar. Ecol. Prog. Ser., **30**:93–104.

Genty, B., Briantais, J.-M., and Barber, N.R. (1989) The relationship between the quantum yield of photosynthetic electron transport and quenching of chlorophyll fluorescence. Biochimica et Biophysica Acta **990**:87–92.

Gerard, V.A. (1986) Photosynthetic characteristics of giant kelp (*Macrocystis pyrifera*) determined *in situ*. Mar. Biol. **90**:473–482.

Gerard, V.A. (1988) Ecotypic differentiation in light-related traits of the kelp *Laminaria saccharina*. Mar. Biol. **97**:25–36.

Gieskes, W.C.C. and Kraay, G.W. (1986) Floristic and physiological differences between the shallow and the deep nanophytoplankton community in the euphotic zone of the open tropical Atlantic revealed by HPLC analysis of pigments. Mar. Biol. **91**:567–576.

Gieskes, W.W.C. and Kraay, G.W. (1989) Estimating the carbon-specific growth rate of the major algal species groups in eastern Indonesian waters by ^{14}C labeling of taxon-specific carotenoids. Deep-Sea Res., **36**:1127–1139.

Gieskes, W.W.C., Kraay, G.W., Nontji, A., Setiapermana, D., and Sutomo (1988) Monsoonal alteration of a mixed and a layered structure in the phytoplankton of the euphotic zone of the Banda Sea (Indonesia): a mathematical analysis of algal pigment fingerprints. Net. J. Sea Res. **22**:123–137.

Glibert, P.M., Kana, T.M., Olson, R.J., Kirchman, D.L., and Alberte, R.S. (1986) Clonal comparisons of growth and photosynthetic responses to nitrogen availability in marine *Synechococcus* spp. J. Exp. Mar. Biol. Ecol. **101**:199–208.

Glidewell, S.M., and Raven, J.A. (1975) Measurement of simultaneous oxygen evolution and uptake in *Hydrodictyon africanum*. J. Exp. Bot. **26**:479–488.

Glover, H.E. (1985) The physiology and ecology of the marine cyanobacterial genus *Synechococcus*. Adv. Aquat. Microbiol. **3**:49–107.

Göbel, F. (1978a) Quantum efficiencies of growth. *In*: The Photosynthetic Bacteria, R.K. Clayton and W.S. Sistrom (Eds.), Plenum Press, New York, pp. 907–925.

Göbel, F. (1978b) Direct measurement of pure absorbance spectra of living phototrophic microorganisms. Biochim. Biophys. Acta **538**:593–602.

Goldman, J.C. (1980) Physiological processes, nutrient availability, and the concept of relative growth rate in marine phytoplankton ecology. *In* Brookhaven Symposium Biology, Vol. 31, Plenum, New York, pp. 179–194.

Goldman, J.C., and Brewer, P.G. (1980) Effect of nitrogen source and growth rate on phytoplankton-mediated changes in alkalinity. Limnol. Oceanogr. **25**:352–357.

Goldman, J.C., and Dennett, M.R. (1985) Susceptibility of some marine phytoplankton species to cell breakage during filtration and post-filtration rinsing. J. Exp. Mar. Biol. Ecol. **86**:47–58.

Goldman, J.C., and Dennett, M.R. (1986) Dark CO_2 fixation by the diatom *Chaetoceros simplex* in response to nitrogen pulsing. Mar. Biol. **90**:493–500.

Goldman, J.C., McCarthy, J.J., and Peavey, D.G. (1979) Growth rate influence on the chemical composition of phytoplankton in oceanic waters. Nature **279**:210–215.

Goodwin, T.W. (1980) The Biochemistry of the Carotenoids, Vol. 1, Plants, Chapman and Hall, London.

Gordon, H.R., and Morel, A.Y. (1983) Lecture Notes on Coastal and Estuarine Studies, Vol. 4, Remote Assessment of Ocean Color for Interpretation of Satellite Visible Imagery: A Review. Springer-Verlag, New York.

Gordon, H.R. Brown, O.B., and Jacobs. M.M. (1975) Computed relationships between the inherent and apparent optical properties of a flat homogeneous ocean. Appl. Opt. **14**:417–427.

Gordon, H.R., (1983) Phytoplankton pigment concentrations in the Middle Atlantic Bight: comparison of ship determinations and CZCS estimates. Appl. Opt. **22**:20–36.

Govindjee, and Satoh, K. (1986) Fluorescence properties of chlorophyll *b*- and chloropyll *c*-containing algae. *In* Light Emission by Plants and Bacteria, Govindjee, J. Amesz and D.C. Fork (Eds.), Academic Press, Orlando, pp. 4497–4537.

Govindjee, Amesz, J., and Fork, D.C. (1986) Light Emission by Plants and Bacteria, Academic Press, Orlando.

Gower, J.F.R. (1980) Observations of *in situ* fluorescence of chlorophyll *a* in Saanich Inlet. Boundary-Layer Meteorol. **18**:235–245.

Graham, D., and Smillie, R.M. (1971) Chloroplasts (and lamellae): algal preparations. *In* Methods in Enzymology, A. San Pietro (Ed.), Academic Press, New York, pp. 228–242.

Grande, K.D., Marra, J., Langdon, C., Heinemann, K., and Bender, M.L. (1989) Rates of respiration in the light measured in marine phytoplankton using an ^{18}O isotope-labelling technique. J. Exp. Mar. Biol. Ecol. **129**:95–120.

Grasshoff, K. (1981) The electrochemical determination of oxygen. *In* Marine Electrochemistry. A Practical Introduction, M. Whitfield and D. Jager (Eds.), John Wiley, New York, pp. 327–420.

Grasshoff, K., Ehrhardt, M., and Kremling, K. (1983) Methods of Seawater Analysis, Verlag Chemie, Deerfield Beach FL.

Gregory, R.P.F. (1989) Biochemistry of Photosynthesis, John Wiley, New York.

Griffith, P.C., Cubit, J.D., Adey, W.H., and Norris, J.N. (1987) Computer-automated flow respirometry: metabolism measurements on a Caribbean reef flat and in a microcosm. Limnol. Oceanogr. **32**:442–451.

Griffiths, D.J. (1973) Factors affecting the photosynthetic capacity of laboratory cultures of the diatom *Phaeodactylum tricornutum*. Mar. Biol. **21**:91–97.

Grum, E., and Luckey, G.W. (1968) Optical sphere paint and a working standard of reflectance. Appl. Opt. **7**:2289–2294.

Guillard, R.R.L., and Hellebust, J.A. (1971) Growth and production of extracellular substances by two strains of *Phaeocystis poucheti*. J. Phycol. **7**:330–338.

Gust, G. (1977) Turbulence and waves inside flexible-wall systems designed for biological studies. Mar. Biol. **42**:47–53.

Gust, G., Booij, K., Helder, W., and Sundby, B. (1987) On the velocity sensitivity (stirring effect) of polarographic oxygen microelectrodes. Neth. J. Sea Res. **21**:255–263.

Gutschick, V.P. (1984) Photosynthesis model for C_3 leaves incorporating CO_2 transport, propagation of radiation, and biochemistry. 1. Kinetics and their parameterization. Photosynthetica **18**:549–568.

Haardt, H., and Maske, H. (1987) Specific *in vivo* absorption coefficient of chlorophyll *a* at 675 nm. Limnol. Oceanogr. **32**:608–619.

Haldane, J.B.S. (1930) Enzymes, Longmans Green, London.

Halldal, P. (1964) Ultraviolet action spectra of photosynthesis and photosynthetic inhibition in a green and a red alga. Physiol. Plant, **17**:414–421.

Hallegraeff, G.M., and Jeffrey, S.W. (1985) Description of new chlorophyll *a* alteration products in marine phytoplankton. Deep-Sea Res. **32**:697–705.

Hama, T. (1988) [13] C-GC-MS analysis of photosynthetic products of phytoplankton population in the regional upwelling around the Izu Islands, Japan. Deep-Sea Res. **35**:91–110.

Hama, T., and Handa, N. (1987) Pattern of organic production by natural phytoplankton population in a eutrophic lake. 2. Extracellular products. Arch. Hydrobiol. **109**:227–243.

Hama, T., Miyazaki, T., Ogawa, Y., Iwakuma, T., Takahashi, M., Otsuki, A., and Ichimura, S. (1983) Measurement of photosynthetic production of a marine phytoplankton population using a stable ^{13}C isotope. Mar. Biol. **73**:31–36.

Hama, T., Handa, N., and Hama, J. (1987) Determination of amino acid production rate of a marine phytoplankton population with ^{13}C and gas chromatography-mass spectrometry. Limnol. Oceanogr. **33**:1144–1153.

Hama, T., Matsunaga, K., Handa, N., and Takahashi, M. (1988) Day-night changes in production of carbohydrate and protein by natural phytoplankton population from Lake Biwa, Japan. J. Plankton Res., **10**:941–955.

Hansson, I. (1973) A new set of pH-scales and standard buffers for sea water. Deep-Sea Res. **20**:479–491.

Hansson, O., and Wydrzynski, T. (1990) Current perceptions of photosystem II. Photosyn. Res. **23**:131–162.

Harding, L.W. Jr., Prezelin, B.B., and Sweeney, B.M. (1981) Diel periodicity of photosynthesis in marine phytoplankton. Mar. Biol., **61**:95–105.

Harding, L.W. Jr., Prezelin, B.B., Sweeney, B.M. and Cox, J.L. (1982) Diel oscillations of the photosynthesis-irradiance (PI) relationship in natural assemblages of phytoplankton. Mar. Biol. **67**:167–178.

Harding, L.W. Jr., Meeson, B.W., and Fisher, T.R. Jr. (1986) Phytoplankton production in two east coast estuaries: photosynthesis-light functions and patterns of carbon assimilation in Chesapeake and Delaware Bays. Est. Coast. Shelf Sci. **23**:773–806.

Harding, L.W. Jr., Fisher, T.R. Jr., and Tyler, M.A. (1987) Adaptive responses of photosynthesis in phytoplankton: specificity to time-scale of change in light. Biol. Oceanogr. **4**:403–437.

Harlin, M.M., and Craigie, J.S. (1975) The distribution of photosynthate in *Ascophyllum nodosum* as it relates to epiphytic *Polysiphonia lanosa*. J. Phycol. **11**:109–113.

Harris, G.P. (1973) Diel and annual cycles of net plankton photosynthesis in Lake Ontario., Fish Res. Bd. Can. J. **30**:1779–1787.

Harris, G.P. (1978) Photosynthesis, productivity and growth: the physiological ecology of phytoplankton. Arch Hydrobiol. **10**:1–171.

Harris, G.P., and Lott, J.N.A. (1973) Light intensity and photosynthetic rates in phytoplankton. J. Fish Res. Bd. Can. **30**:1771–1778.

Harris, G.P., and Piccinin, B.B. (1977) Photosynthesis by natural phytoplankton populations. Arch. Hydrobiol. **80**:405–457.

Harris, G.P., Haffner, G.D., and Piccinin, B.B. (1980) Physical variability and phytoplankton communities. II. Primary production of phytoplankton in a physically variable environment. Arch. Hydrobiol. **88**:393–425.

Harrison, W.G., and Platt, T. (1980) Variations in assimilation number of coastal marine phytoplankton: effects of environmental co-variates. J. Plankton Res. **2**:249–260.

Hartman, B., and Hammond, D.E. (1985) Gas exchange in San Francisco Bay. Hydrobiologia **129**:59–68.

Hatcher, B.G. (1977) An apparatus for measuring photosynthesis and respiration of intact large marine algae and comparison of results with those from experiments with tissue segments. Mar. Biol. **43**:381–385.

Hatcher, B.G., Chapman, A.R.O., and Mann, K.H. (1977) An annual carbon budget for the kelp *Laminania longienins*. Mar. Biol. **44**:85–96.

Haxo, F.T. (1985) Photosynthetic action spectrum of the coccolithophorid *Emiliania huxlevi* (Haptophyceae): 19′hexanoyloxyfucoxanthin as antenna pigment. J. Phycol. **21**:282–287.

Haxo, F.T., and Blinks, L.R. (1950) Photosynthetic action spectra of marine algae. J. Gen. Physiol. **33**:389–422.

Hayward, T.L., Venrick, E.L., and McGowan, J.A. (1983) Environmental heterogeneity in the central North Pacific. J. Mar. Res. **41**:711–729.

Hayward, T.L., and Venrick, E.L. (1982) Relation between surface chlorophyll, integrated chlorophyll and integrated primary production. Mar. Biol. **69**:247–252.

Heber, U., Takahama, U., Neimanis, S., and Shimizu-Takahama, M. (1982) Transport as the basis of the Kok effect. Levels of some photosynthetic intermediates and activation of light-regulated enzymes during photosynthesis of chloroplasts and green leaf protoplasts. Biochim. Biophys. Acta **679**:287–299.

Hecky, R.E., and Fee, E.J. (1981) Primary production and rates of algal growth in Lake Tanganyika. Limnol. Oceanogr. 26:532–547.

Hellebust, J.A. (1965) Excretion of some organic compounds by marine phytoplankton. Limnol. Oceanogr. 10:192–206.

Herczeg, A.L., and Hesslein, R.H. (1984) Determination of hydrogen ion concentration in softwater lakes using carbon dioxide equilibria. Geochim. Cosmochim. Acta 48:837–845.

Herczeg, A. L., Broecker, W.S., Anderson, R.F., Schiff, S.L., and Schindler, D.W. (1985) A new method for monitoring temporal trends in the acidity of fresh waters. Nature 315:133–135.

Herzig, R. and Falkowski, P.G. (1989) Nitrogen limitation in *Isochrysis galbana* (Haptophyceae). I. Photosynthetic energy conversion and growth efficiencies. J. Phycol., 25:462–471.

Hipkins, M.F. (1978) The emission yields of delayed and prompt fluoresence from chloroplasts. Biochim. Biophys. Acta 502:161–168.

Hipkins, M.F., and Baker, N.R. (1986) Photosynthesis Energy Transduction. A Practical Approach, IRL Press, Oxford.

Hitchcock, G.L. (1982) A comparative study of size-dependent organic composition of marine diatoms and dinoflagellates. J. Plankton Res. 6:219–237.

Hitchcock, G.L. (1983) Photosynthate partitioning in cultured marine phytoplankton. I. Dinoflagellates. J. Exp. Mar. Biol. Ecol. 69:21–36.

Hitchman, M.L. (1978) Measurement of Dissolved Oxygen. John Wiley, New York.

Hobson, L.A., Morris, W.J., and Pirquet, K.T. (1976) Theoretical and experimental analysis of the ^{14}C technique and its use in studies of primary production. J. Fish. Res. Bd. Can. 33:1715–1721.

Hoch, G., Owens, O.V.H., and Kok, B. (1963) Photosynthesis and respiration. Arch. Biochem. Biophys. 101:171–180.

Hodges, M., and Barber, J. (1983) The significance of the kinetic analysis of fluorescence induction in DCMU-inhibited chloroplasts in terms of photosystem 2 connectivity and heterogeneity. Fed. Eur. Biochem. Soc. Lett. 160:177–181.

Hodges, M., Cornic, G., and Briantais, J.-M. (1989) Chlorophyll fluorescence from spinach leaves: resolution of non-photochemical quenching. Biochim. Biophys. Acta 974:289–293.

Hoffman, W.E., and Dawes, C.J. (1980) Photosynthetic rates and primary production by two Florida benthic red algal species from a salt marsh and a mangrove community. Bull. Mar. Sci. 30:358–364.

Hofslagare, O., Samuelsson, G., Hällgren, J.-E., Pejryd, C., and Sjöberg, S. (1985) A comparison between three methods of measuring photosynthetic uptake of inorganic carbon in algae. Photosynthetica 19:578–585.

Hoge, F.E., and Swift, R.N. (1981) Airborne simultaneous spectroscopic detection of laser-induced water Raman backscatter and fluorescence from chlorophyll *a* and other naturally occurring pigments. Appl. Opt. 20:3197–3205.

Hoge, F.E., Berry, R.E., and Swift, R.N. (1986a) Active-passive airborne ocean color measurement. 1. Instrumentation. Appl. Opt. 25:39–47.

Hoge, F.E., Swift, R.N., and Yungel, J.K. (1986b) Active-passive airborne ocean color measurement. 2. Applications. Appl. Opt. **25**:48–57.

Holligan, P.M., Viollier, M., Dupouy, C., and Aiken, J. (1983) Satellite studies on the distributions of chlorophyll and dinoflagellate blooms in the western English Channel. Cont. Shelf Res. **2**:81–96.

Holmes, J.J., Weger, H.G., and Turpin, D.H. (1989) Chlorophyll *a* fluorescence predicts total photosynthetic electron flow to CO_2 or NO_3^- / NO_2^- under transient conditions. Plant Physiol. **91**:331–337.

Holm-Hansen, O., Lorenzen, C.J., Holmes, R.N., and Strickland, J.D.H. (1965) Fluorometric determination of chlorophyll. J. Cons. Perm. Int. Explor. Mer., **30**:3–15.

Hooks, C.E., Bidigare, R.R., Keller, M.D., and Guillard, R.R.L. (1988) Coccoid eukaryotic marine ultraplankters with four different HPLC pigment signatures. J. Phycol. **24**:571–580.

Horton, P. (1985) Interactions between electron transfer and carbon assimilation. *In* Photosynthetic Mechanisms and the Environment, J. Barber and N.R. Baker (Eds.), Elsevier, Amsterdam, pp. 135–187.

Horton, P., and Croze, E. (1979) Characterization of two quenchers of chlorophyll fluorescence with different midpoint oxidation—reduction potentials in chloroplasts. Biochim. Biophys. Acta **545**:188–201.

Horton, P., and Hague, A. (1988) Studies on the induction of chlorophyll fluorescence in isolated barley protoplasts. IV. Resolution of nonphotochemical quenching. Biochim. Biophys. Acta **931**:107–115.

Hsu, B.-D., Lee, Y.-S., and Rang, Y.-R. (1989) A method for analysis of fluorescence induction curve from DCMU-poisoned chloroplasts. Biochim. Biophys. Acta **975**:44–49.

Hulst, van de H.C. (1957) Light Scattering by Small Particles. John Wiley, New York.

Idle, D.B., and Proctor, C.W. (1983) An integrating sphere leaf chamber. Plant Cell Environ. **6**:437–439.

Inamura, I., Ochiai, H., Toki, K., Watanabe, S., Hikino, S., and Araki, T. (1983) Preparation and properties of chlorophyll/water-soluble macromolecular complexes in water. Stabilization of chlorophyll aggregates in the water-soluble macromolecule. Photochem. Photobiol. **38**:37–44.

Ireland, C.R., Long, S.P., and Baker, N.P. (1984) The relationship between carbon dioxide fixation and chlorophyll *a* fluorescence during induction of photosynthesis in maize leaves at different temperatures and carbon dioxide concentrations. Planta **160**:550–558.

Ittekkot, V., Brockman, U., Michaelis, W., and Degens, E.T. (1981) Dissolved free and combined carbohydrates during a phytoplankton bloom in the Northern North Sea. Mar. Ecol. Prog. Ser. **4**:299–305.

Itturiaga, R., and Marra, J. (1988) Temporal and spatial variability of chrococcoid cyanobacteria *Synechococcus* spp. specific growth rates and their contribution to primary production in the Sargasso Sea. Mar. Ecol. Prog. Ser. **44**:175–181.

Iturriaga, R., and Zsolnay, A. (1983) Transformation of some dissolved organic compounds by a natural heterotrophic population. Mar. Biol. **62**:125–129.

Itturiaga, R., Mitchell, B.G., & Kiefer, D.A. (1988) Microphotometric analysis of individual particle absorption spectra. Limnol. Oceanogr., **33**:128–134.

Iverson, R.L., and Curl, H. (1973) Action spectrum of photosynthesis for *Skeletonema costatum* obtained with carbon-14. Physiol. Plant **28**:498–502.

Jackson, G.A. (1983) Zooplankton grazing effects on ^{14}C-based phytoplankton production measurements: a theoretical study. J. Plankton Res. **5**:83–94.

Jacquez, J.A., and Kuppenhelm, H.F. (1955) Theory of the integrating sphere. J. Opt. Soc. Am. **45**:460–470.

Jassby, A.D. (1978) Polarographic measurements of photosynthesis and respiration. In Handbook of Phycological Methods, Physiological and Biochemical Methods, J.A. Hellebust and J.S. Craigie (Eds.), Cambridge University Press, New York, pp. 285–296.

Jassby, A.D., and Platt, T. (1976) Mathematical formulation of the relationship between photosynthesis and light for phytoplankton. Limnol. Oceanogr. **21**:540–547.

Jeffrey, S.W. (1968) Quantitative thin-layer chromatography of chlorophylls and carotenoids from marine algae. Biochim. Biophys. Acta **162**:271–285.

Jeffrey, S.W. (1972) Preparation and some properties of chlorophylls c_1 and c_2 from marine algae. Biochim. Biophys. Acta **279**:15–33.

Jeffrey, S.W. (1974) Profiles of photosynthetic pigments in the ocean using thin layer chromatography. Mar. Biol. **26**:101–110.

Jeffrey, S.W. (1976) A report of green algal pigments in the Central North Pacific Ocean. Mar. Biol. **37**:33–37.

Jeffrey, S.W. (1980) Algal pigment systems. *In* Primary Productivity in the Sea, P.G. Falkowski (Ed.), Plenum, New York, pp. 33–58.

Jeffrey, S.W. (1981) An improved thin-layer chromatography technique for marine phytoplankton pigments. Limnol. Oceanogr. **26**:191–197.

Jeffrey, S.W., and Hallegraeff, G.M. (1987a) Chlorophyllase distribution in ten classes of phytoplankton: a problem for chlorophyll analysis. Mar. Ecol. Prog. Ser. **35**:293–304.

Jeffrey, S.W., and Hallegraeff, GM. (1987b) Phytoplankton pigments, species and light climate in a complex warm-core eddy of the East Australian Current. Deep-Sea Res. **34**:649–673.

Jeffrey, S.W., and Haxo, F.T. (1968) Photosynthetic pigments of symbiotic dinoflagellates (zooxanthellae) from corals and clams. Biol. Bull. **135**:149–165.

Jeffrey, S.W., and Humphrey, G.F. (1975) New spectrophotometric equations for determining chlorophylls a, b, c_1 and c_2 in higher plants, algae and natural phytoplankton. Biochem. Physiol. Pflanz. **167**:191–194.

Jenkins, F.A., and White, H.E. (1976) Fundamentals of Optics, McGraw-Hill, Kogakusha, Ltd., Tokyo, pp. 746.

Jenkins, W.J., and Goldman, J.C. (1985) Seasonal oxygen cycling and primary production in the Sargasso Sea. J. Mar. Res. **43**:465–491.

Jensen, P.R., Gibson, R.A., Littler, M.M., and Littler, D.S. (1985) Photosynthesis and calcification in four deep-water *Halimeda* species (Chlorophyceae, Caulerpales). Deep-Sea Res. **32**:451–464.

Jensen, A. (1978) Chlorophylls and carotenoids. In: J.A. Hellebust and J.S. Craigie (Eds), Handbook of Phycological Methods: Physiological and Biochemical Methods. Cambridge University Press, NY, pp. 60–70.

Jenson, L.M. (1983) Phytoplankton release of extracellular organic carbon, molecular weight composition, and bacterial assimilation. Mar. Ecol. Prog. Ser. **11**:39–48.

Jerlov, N.G. (1968) Optical Oceanography, Elsevier, Amsterdam.

Jewson, D.H. (1976) The interaction of components controlling net photosynthesis in a well-mixed lake (Lough Neagh, Northern Ireland). Freshwater Biol. **6**:551–576.

Jewson, D.H. (1977) A comparison between *in situ* photosynthetic rates determined using ^{14}C uptake and oxygen evolution methods in Lough Neagh, Northern Ireland. Proc. R. Irish Acad. **B77**:87–99.

Jewson, D.H., Talling, J.F., Dring, M.J., Tilzer, H.H., Heaney, S.I., and Cunningham, C. (1984) Measurement of photosynthetically available radiation in freshwater: comparative tests of some current instruments used in studies of primary production. J. Plankton Res. **6**:259–273.

Johnson, K.S., Pytkowicz, R.M., and Wong, C.S. (1979) Biological production and the exchange of oxygen and carbon dioxide across the sea surface in Stuart Channel, British Columbia. Limnol. Oceanogr. **24**:474–482.

Johnson, K.M., Burney, C.M., and Sieburth, J. McN. (1981) Enigmatic marine ecosystem metabolism measured by direct diel CO_2 and O_2 flux in conjunction with DOC release and uptake. Mar. Biol. **65**:49–60.

Johnson, K.M., King, A.E., and Sieburth, J.McN. (1985) Coulometric TCO_2 analyses for marine studies: an introduction. Mar. Chem. **16**:61–82.

Johnson, K.M., Sieburth, J. McN., Williams, P.J.LeB., and Brandstrom, L. (1987) Coulometric total carbon dioxide analysis for marine studies: automation and calibration. Mar. Chem. **21**:117–133.

Johnson, W.S., Gigon, A., Gulmon, S.L., and Mooney, H.A. (1974) Comparative photosynthetic capacities of intertidal algae under exposed and submerged conditions. Ecology **55**:240–453.

Johnston, A.M. and Raven, J.A. (1989) Extraction, partial purification and characterization of phosphoenolpyruvate carboxykinase from *Ascophyllum nodosum* (Phaeophyceae) J. Phycol. **25**:568–576.

Joliot, P., and Joliot, A. (1968) A polarographic method for detection of oxygen production and reduction of Hill reagent by isolated chloroplasts. Biochim. Biophys. Acta **153**:625–634.

Jones, J.G. (1974) A method for the observation and enumeration of epibenthic algae directly on the surface of stone. Oecologia (Berlin) **16**:1–8.

Jorgensen, B.B., Cohen, Y., and Des Marias, D.J. (1987) Photosynthetic action spectra and adaptation to spectral light distribution in a benthic cyanobacteria mat. Appl. Environ. Microbiol. **53**:879–886.

Jorgensen, B.B., and Des Marias, D.J. (1988) Optical properties of benthic photosynthetic communities: fiber-optic studies of cyanobacterial mats. Limnol. Oceanogr. **33**:99–113.

Juday, C., Blair, J.M., and Wilda, E.F. (1943) The photosynthetic activities of the aquatic plants of Little John Lake, Vilas County, Wisconsin. Am. Midl. Nat., 30:426–446.

Kairesalo, T., Gunnarsson, K., Jonsson, G. St., and Jonasson, P.M. (1987) The occurrence and photosynthetic activity on the tips of *Nitella opaca* Ag. (Charophyceae). Aquat. Bot. 28:333–340.

Kana, T.M. and Glibert, P.M. (1987) Effect of irradiances up to 2000 μE m^{-2} s^{-1} on marine *Synechococcus* WH7803. I. Growth, pigmentation, and cell composition. Deep-Sea Res., 34:479–495.

Kanda, J., and Hattori, A. (1988) Ammonium uptake and synthesis of cellular nitrogenous macromolecules in phytoplankton. Limnol. Oceanogr. 33:1568–1579.

Kanda, J., Saino, T., and Hattori, A. (1988). Nitrogen nutrition and physiological state of natural populations of phytoplankton in surface waters of the western Pacific Ocean. Limnol. Oceanogr. 33:1580–1585.

Kanwisher, J.W. (1966) Photosynthesis and respiration in some seaweeds. *In* Some Contemporary Studies in Marine Science, H. Barnes (Ed.), Hafner Publishing Company, New York, pp. 407–420.

Kaplan, A., and Berry, J.A. (1981) Glycolate excretion and the oxygen to carbon dioxide net exchange ratio during photosynthesis in *Chlamydomonas reinhardtii*. Plant Physiol. 67:229–232.

Kato, K., and Stabel, H. (1984) Studies on the carbon flux from phyto- to bacterioplankton communities in Lake Constance. Arch. Hydrobiol. 102:177–192.

Kautsky, H., and Hirsch, A. (1931) Neue versuche zur kohlenstoffassimilation. Naturwissenschaften 19:964.

Keiller, D.R., and Walker, D.A. (1990). The use of chlorophyll fluorescence to predict CO_2 fixation during photosynthetic oscillations. *Proc. R. Soc. London,* B241:59–64.

Keller, A.A. (1988a) An empirical model of primary productivity (^{14}C) using mesocosm data along a nutrient gradient. J. Plankton Res. 10:813–834.

Keller, A.A. (1988b) Estimating phytoplankton productivity from light availability and biomass in the MERL microcosms and Naragansett Bay. Mar. Ecol. Prog. Ser. 45:159–168.

Kemp, W.M. and Boynton, W.R. (1980) Influence of biological and physical processes on dissolved oxygen dynamics in an estuarine system: implications for measurement of community metabolism. Estuarine Coastal Shelf Sci., 11:407–431.

Khailov, K.M., and Burlakova, Z.P. (1969) Release of dissolved organic matter by marine seaweeds and distribution of their total organic production to inshore communities. Limnol. Ocean. 14:521–527.

Kieber, D.J., and Mopper, K. (1987) Photochemical formation of glyoxylic and pyruvic acids in seawater. Mar. Chem. 21:135–149.

Kiefer, D.A. (1973) Chlorophyll *a* fluorescence in marine centric diatoms: responses of chloroplasts to light and nutrient stress. Mar. Biol. 23:39–46.

Kiefer, D.A., and Mitchell, B.G. (1983) A simple, steady state description of phytoplankton growth based on absorption cross section and quantum efficiency. Limnol. Oceanogr. 28:770–776.

Kiefer, D.A., and SooHoo, J.B. (1982) Spectral absorption by marine particles of coastal waters of Baja California. Limnol. Oceanogr. 27:492–499.

Kiefer, D.A., Chamberlin, W.S., and Booth, C.R. (1989) Natural fluorescence of chlorophyll *a*: relationship to photosynthesis and chlorophyll concentration in the western South Pacific gyre. Limnol. Oceanogr. 34:860–881.

Kinsey, D.W.K. (1985) Open-flow systems. *In* Handbook of Phycological Methods, Vol. 4, Ecological Field Methods: Macroalgae, M.M. Littler and D.S. Littler (Eds.), Cambridge University Press, Cambridge, pp. 427–460.

Kirk, J.T.O. (1983) Light and Photosynthesis in Aquatic Ecosystems. Cambridge University Press, Cambridge.

Klimov, V.V., and Krasnovskii, A.A. (1981) Pheaophytin as the primary electron acceptor in photosystem 2 reaction centres. Photosynthetica 15:592–609.

Knoechel, R., and Kalff, J. (1976a) The applicability of grain density autoradiography to the quantitative determination of algal species production: a critique. Limnol. Oceanogr. 21:583–590.

Knoechel, R., and Kalff, J. (1976b) Track autoradiography: a method for the determination of phytoplankton species productivity. Limnol. Oceanogr. 21:590–596.

Knoechel, R., and Kalff, J. (1978) An *in situ* study of the productivity and population dynamics of five freshwater planktonic diatom species. Limnol. Oceanogr. 23:195–218.

Kok, B. (1948) A critical consideration of the quantum yield of *Chlorella* photosynthesis. Enzymologia 13:1–56.

Kok, B. (1949) On the interrelation of respiration and photosynthesis in green plants. Biochim. Biophys. Acta 3:625–631.

Kok, B. (1951) Photo-induced interactions in metabolism of green plant cells. Symp. Soc. Exp. Biol. 5:211–221.

Kok, B. (1952) On the efficiency of *Chlorella* growth. Acta Bot. Neerl. 1:445–467.

Kok, B. (1956) On the inhibition of photosynthesis by intense light. Biochim. Biophys. Acta 21:234–244.

Kolber, Z., Wyman, K.O., and Falkowski, P.G. (1990) Natural variability in photosynthetic energy conversion efficiency: a field study in the Gulf of Maine. Limnol. Oceanogr. 35:72–79.

Krause, G.H., and Somersalo, S. (1989) Fluorescence as a tool in photosynthesis research: application in studies of photoinhibition, cold acclimation and freezing stress. Phil. Trans. R. Soc. London B323:281–293.

Krause, G.H., and Weis, E. (1984) Chlorophyll fluorescence as a tool in plant physiology. II. Interpretation of fluorescence signals. Photosynth. Res. 5:139–157.

Krause, G.H., Vernotte, C., and Briantais, J.-M. (1982) Photoinduced quenching of chlorophyll fluorescence in intact chloroplast and algae. Resolution into two components. Biochim. Biophys. Acta 679:116–124.

Krawiec, R.W. (1982) Autecology and clonal variability of the marine centric diatom *Thalassiosira rotula* (Bacillariophyceae) in response to light, temperature and salinity. Mar. Biol. 69:79–90.

Kremer, B.P. (1981) C_4-metabolism in marine brown macrophytic algae. Z. Naturforsch **36**:840–847.

Kuppers, U., and Kremer, B.P. (1978) Longitudinal profiles of CO_2-fixation capacities in marine macroalgae. Plant Physiol. **62**:49–54.

Kursar, T.A., and Alberte, R.S. (1983) Photosynthetic unit organization in a red alga. Plant Physiol. **72**:409–414.

Laanbroek, H.J., Verplanke, J.C., de Visscher, P.R.M., and de Vuyst, R. (1985) Distribution of phyto- and bacterioplankton growth and biomass parameters, dissolved inorganic nutrients and free amino acids during a spring bloom in the Oosterschelde Basin, The Netherlands. Mar. Ecol. Prog. Ser. **25**:1–11.

Lancelot, C. (1984) Extracellular release of small and large molecules by phytoplankton in the southern bight of the North Sea. Est. Coast. Shelf Sci. **18**:65–77.

Landry, M.R., and Hassett, R.P. (1982) Estimating the grazing impact of marine microzooplankton. Mar. Biol. **67**:283–288.

Langdon, C. (1984) Dissolved oxygen monitoring system using a pulsed electrode: design, performance and evaluation. Deep-Sea Res. **31**:1357–1367.

Langdon, C. (1988) On the causes of interspecific differences in the growth-irradiance relationship for phytoplankton. II. A general review. J. Plank. Res. **10**:1291–1312.

Lanne, R.W.P.M., Gieskes, W.W.C., Kraay, G.W., and Eversdijk, A. (1985) Oxygen consumption from natural waters by photo-oxidizing processes. Neth. J. Sea Res. **19**:125–128.

Lapointe, B.E., and Duke, C.S. (1984) Biochemical strategies for growth of *Gracilaria tikvahiae* (Rhodophyta) in relation to light intensity and nitrogen availability. J. Phycol. **20**:488–495.

Larkum, A.W.D., and Barrett, J. (1983) Light harvesting process in algae. Adv. Bot. Res. **10**:1–219.

Latimer, P., Bannister, T.T., and Rabinowitch, E. (1956) Quantum yields of fluorescence of plant pigments. Science **124**:585–586.

Lavorel, J., Breton, J., and Lutz, M. (1986) Methodological principles of measurement of light emitted by photosynthetic systems. *In* Light Emission by Plants and Bacteria, Govindjee, J. Amesz, and D.C. Fork (Eds.) Academic Press, Orlando, pp. 57–98.

Laws, E.A., and Bannister, T.T. (1980) Nutrient- and light-limited growth of *Thalassiosira fluviatilis* in continuous culture, with implications for phytoplankton growth in the ocean. Limnol. Oceanogr. **25**:457–473.

Laws, E.A. Redalje, D.G., Haas, L.W. Bienfang, P.K., Eppley, R.W., Harrison, W.G., Karl, D.H., and Marra, J. (1984) High phytoplankton growth and production rates in oligotrophic Hawaiian coastal waters. Limnol. Oceanogr. **29**:1161–1169.

Laws, E.A., Jones, D.R., Terry, K.L., and Hirata, J.A. (1985) Modifications in recent models of phytoplankton growth: theoretical developments and experimental examination of predictions. J. Theor. Biol. **114**:323–341.

Laws, E.A., Ditullio, G.R., and Redalje, D.G. (1987) High phytoplankton growth and production rates in the North Pacific subtropical gyre. Limnol. Oceanogr. **32**:905–918.

Lederman, T.C., and Tett, P. (1981) Problems in modelling the photosynthesis-light relationship for phytoplankton. Bot. Mar. **24**:125–134.

Lee, A.C., and Butler, W.L. (1976) The efficiency of energy transfer from photosystem II to photosystem I in *Porphyridium cruentum*. Proc. Natl. Acad. Sci. USA **73**:3957–3960.

Lee, Y.-K., Tan, H.-M., and Hew, C.-S. (1985) The effect of growth temperature on the bioenergetics of photosynthetic algal cultures. Biotech. Bioengin. **27**:555–561.

Legendre, L., Demers, S., Yentsch, C.M., and Yentsch, C.S. (1983) The ^{14}C method: patterns of dark CO_2 fixation and DCMU correction to replace the dark bottle. Limnol. Oceanogr. **28**:996–1003.

Lessard, E.J., and Swift, E. (1986) Dinoflagellates from the North Atlantic classified as phototrophic or heterotrophic by fluorescence microscopy. J. Plankton Res. **8**:1209–1215.

Leverenz, J.W. (1987) Chlorophyll content and light response of shade-adapted conifer needles. Physiol. Plant. **71**:20–29.

Leverenz, J.W. (1988) The effects of illumination sequence, CO_2 concentration, temperature and acclimation on the convexity of the photosynthetic light response curve. Physiol. Plant. **74**:332–341.

Levy, I., and Gantt, E. (1988) Light acclimation in *Porphyridium purpureum* (Rhodophyta): growth, photosynthesis and phycobilisomes. J. Phycol. **24**:452–458.

Lewis, M.R., and Smith, J.C. (1983) A small volume, short-incubation time method for measurement of photosynthesis as a function of incident irradiance. Mar. Ecol. Prog. Ser. **13**:99–102.

Lewis, M.R., Kemp, W.M., Cunningham, J.J., and Stevenson, J.C. (1982) A rapid technique for preparation of aquatic macrophyte samples for measuring ^{14}C incorporation. Aquat. Bot. **13**:203–207.

Lewis, M.R., Warnock, R.E., and Platt, T. (1985) Absorption and photosynthetic action spectra for natural phytoplankton populations: implications for production in the open ocean. Limnol. Oceanogr. **30**:794–806.

Lewis, M.R., Warnock, R.E., and Platt, T. (1986) Photosynthetic response of marine picoplankton at low photon flux. Can. Bull. Fish. Aquat. Sci. **214**:235–250.

Lewis, M.R., Ulloa, O., and Platt, T. (1988) Photosynthetic action, absorption, and quantum yield spectra for a natural population of *Oscillatoria* in the North Atlantic. Limnol. Oceanogr. **33**:92–98.

Ley, A.C. (1986) Relationships among cell chlorophyll content, photosytem II light harvesting and the quantum yield for oxygen production in *Chlorella*. Photosynth. Res. **10**:189–196.

Ley, A.C., and Mauzerall, D.C. (1982) Absolute absorption cross-section for photosystem II and the minimum quantum requirement for photosynthesis in *Chlorella vulgaris*. Biochim. Biophys. Acta **680**:95–106.

Li, W.K.W. (1982) Estimating heterotrophic bacterial productivity by inorganic radiocarbon uptake: importance of establishing time courses of uptake. Mar. Ecol. Prog. Ser. **8**:167–172.

Li, W.K.W. (1986) Experimental approaches to field measurements: methods and interpretation. Can. Bull. Fish. Aquat. Sci. **214**:251–286.

Li, W.K.W., and Harrison, W.G. (1982) Carbon flow into the end-products of photosynthesis in short and long incubations of a natural phytoplankton population. Mar. Biol. **72**:175–182.

Li, W.K.W., and Morris, I. (1982) Temperature adaptation in *Phaeodactylum tricornutum* Bohlin: Photosynthetic rate compensation and capacity. J. Exp. Mar. Biol. Ecol. **58**:135–150.

Li, W.K.W., Glover, H.E., and Morris, I. (1980) Physiology of carbon photoassimilation by *Oscillatoria thiebautii* in the Caribbean Sea. Limnol. Oceanogr. **25**:447–456.

Liaaen-Jensen, S. (1977) Algal carotenoids and chemosystematics. *In* Marine Natural Products Chemistry, D.J. Faulkner, and W.H. Fennical, (Eds.), Plenum Press, New York, pp. 239–259.

Littler, M.M. (1979) The effects of bottle volume, thallus weight, oxygen saturation values, and water movement on apparent photosynthetic rates in marine algae. Aquat. Bot. **7**:21–34.

Littler, M.M., and Arnold, K.E. (1982) Primary productivity of marine macroalgal functional-form groups from southwestern North America. J. Phycol. **18**:307–311.

Littler, M.M., and Littler, D.S. (1985) Deepest known plant life discovered on an uncharted seamount. Science **227**:57–59.

Littler, M.M., Littler, D.S., and Taylor, P.R. (1983) Evolutionary strategies in a tropical barrier reef system: functional-form groups of marine macroalgae. J. Phycol. **19**:229–237.

Loftus, M.E., and Seliger, H.H. (1975) Some limitations of the *in vivo* fluorescence technique. Chesapeake Sci. **16**: 79–92.

Loftus, M.E., Subba Rao, S.V., and Seliger, H.H. (1972) Growth and dissipation of phytoplankton in Chesapeake Bay. I. Response to a large pulse of rainfall. Chesapeake Sci. **13**:282–299.

Lohrenz, S.E., and Taylor, C.D. (1987) Inorganic ^{14}C as a probe of growth rate-dependent variations in intracellular free amino acids and protein composition of NH_4^+-limited continuous cultures of *Nannochloris atomus* Butcher. J. Exp. Mar. Biol. Ecol. **106**:31–55.

Lohrenz, S.E., Arone, R.A., Wiesenburg, D.A., and DePalma, I.P. (1988) Satellite detection of transient enhanced primary production in the western Mediterranean Sea. Nature **335**:245–247.

Lorenzen, C.J. (1970) Surface chlorophyll as an index of the depth, chlorophyll content, and primary production of the euphotic layer. Limnol. Oceanogr. **15**:479–480.

Lorenzen, C.J. (1967) Determination of chlorophyll and pheo-pigments: spectrophotometric equations. Limnol. Oceanogr., **12**:343–346.

Lorenzen, C.J. (1966) A method for continuous measurement of *in vivo* chlorophyll concentration. Deep-Sea Res. **13**:223–247.

Lorenzen, C.J., and Jeffrey, S.W. (1980) Determination of chlorophyll in seawater. UNESCO Tech. Pap. Mar. Sci. **35**:1–20.

Lüning, K., and Dring, M.J. (1985) Action spectra and spectral quantum yield of photosynthesis in marine macroalgae with thin and thick thalli. Mar. Biol. **87**:119–129.

Maberly, S.C., and Spence, D.H.N. (1983) Photosynthetic inorganic carbon use by freshwater plants. J. Ecol. **71**:705–724.

MacCaull, W.A., and Platt, T. (1977) Diel variations in the photosynthetic parameters of coastal marine phytoplankton. Limnol. Oceanogr. **22**:723–731.

Maccio, M., and Langdon, C. (1988) Description of conversion of an EE&G VMCM into a MVMS (multi-variable moored sensor). In Proceedings of the Oceans, 1988 Conference, Baltimore, MD. IEEE, Piscataway, NJ, pp. 1181–1187.

Mague, T.H., Friberg, E., Hughes, D.J., and Morris, I. (1980) Extracellular release of carbon by marine phytoplankton; a physiological approach. Limnol. Oceanogr. **25**:262–279.

Maguire, B., and Neill, W.E. (1971) Species and individual productivity in phytoplankton communities. Ecology **52**:903–907.

Majak, W., Craigie, J.S., and McLaughlan, J. (1966) Photosynthesis in the Rhodophyceae. Can. J. Bot. **44**:541–549.

Malkin, S., and Kok, B. (1966) Fluorescence studies in isolated chloroplasts. I. Number of components involved in the reaction and quantum yields. Biochim. Biophys. Acta **126**:413–432.

Mann, J.E., and Myers, J. (1968) On pigments, growth, and photosynthesis of *Phaeodactylum tricornutum*. J. Phycol. **4**:349–355.

Mann, K.H., Chapman, A.R.O., and Gagné, J.A. (1980) Productivity of seaweeds: the potential and the reality. In Primary Productivity in the Sea, P.J. Falkowski (Ed.), Plenum, New York, pp. 363–380.

Mantoura, R.F.C., and Llewellyn, C.A. (1983) The rapid determination of algal chlorophyll and carotenoid pigments and their breakdown products in natural waters by reverse-phase high-performance liquid chromatography. Anal. Chim. Acta **151**:297–314.

Marra, J. (1978) Phytoplankton photosynthetic response to vertical movement in a mixed layer. Mar. Biol. **46**:203–208.

Marra, J. (1980) Vertical mixing and primary production. In Primary Productivity in the Sea, P.G. Falkowski, (Ed.) Plenum Press, New York, p. 121–137.

Marra, J., Heinemann, K., and Landriau, G. Jr., (1985) Observed and predicted measurements of photosynthesis in a phytoplankton culture exposed to natural irradiance. Mar. Ecol. Prog. Ser. **24**:43–50.

Marra, J., Haas, L.W., and Heinemann, K.R. (1988) Time course of C assimilation and microbial food web. J. Exp. Mar. Biol. Ecol. **115**:263–280.

Marra, J., and Heinemann, K.R. (1984) A comparison between noncontaminating and conventional procedures in primary productivity measurements. Limnol. Oceanogr. **29**:389–392.

Marra, J., and Heinemann, K.R. (1987) Primary production in the North Pacific central gyre: Some new measurements based on ^{14}C. Deep-Sea Res. **34**:1821–1829.

Maske, H., and Haardt, H. (1987) Quantitative *in vivo* absorption spectra of phytoplankton: detrital absorption and comparison with fluorescence excitation spectra. Limnol. Oceanogr. **32**:620–633.

Mauzerall, D. (1972) Light-induced fluorescence changes in *Chlorella* and the primary photoreactions for the production of oxygen. Proc. Nat. Acad. Sci. **69**:1358–1362.

Mauzerall, D. (1977) Porphyrins, chlorophylls and photosynthesis. *In* Encyclopedia of Plant Physiology, Vol. 5A, A. Trebst, and M. Avron, (Eds.), Springer-Verlag, Berlin, pp. 117–124.

Mauzerall, D. (1978) Multiple excitations and the yield of chlorophyll *a* fluorescence in photosynthetic systems. Photochem. Photobiol. **28**:991–998.

Mauzerall, D. (1980) Fluorescence and photosynthesis: gated detection and analysis of nanosecond pulse excitation. Adv. Biol. Med. Phys. **17**:173–198.

Mauzerall, D., and Greenbaum, N.L. (1989) The absolute size of a photosynthetic unit. Biochim. Biophys. Acta **974**:119–140.

McAllister, C.D., Parsons, T.R., and Stephens, K., and Strickland, J.D.H. (1961) Measurements of primary production in coastal sea water using a large-volume plastic sphere. Limnol. Oceanogr. **6**:237–258.

McConville, M.J., Mitchell, C., and Weatherbee, R. (1985) Patterns of carbon assimilation in a microalgal community from annual sea ice. East Antarctica. Polar Biol. **4**:135–141.

McCree, K.J. (1972) Significance of enhancement for calculations based on the action spectrum for photosynthesis. Plant Physiol. **49**:704–706.

McKinley, K.R., and Wetzel, R.G. (1977) Tritium oxide uptake by algae: An independent measure of phytoplankton photosynthesis. Limnol. Oceanogr. **22**:377–380.

McLeod, G.C., and Kanwisher, J. (1962) The quantum efficiency of photosynthesis in ultraviolet light. Physiol. Plant **15**:581–586.

McQuaker, N.R., Kluckner, P.D., and Sandberg, D.K. (1983) Chemical analysis of acid precipitation: pH and acidity determinations. Environ. Sci. Tech. **17**:431–435.

Megard, R.O., Combs, W.S. Jr., Smith, P.D., and Knoll, A.S. (1979) Attenuation of light and daily integrals of photosynthesis attained by planktonic algae. Limnol. Oceanogr. **24**:1038–1050.

Megard, R.O., Tonkyn, D.W., and Senft, W.H. III (1984) Kinetics of oxygenic photosynthesis in planktonic algae. J. Plankton Res. **6**:325–337.

Megard, R.O., Berman, T., Curtis, P.J., and Vaughan, P.W. (1985) Dependence of phytoplankton assimilation quotients on light and nitrogen source: implications for oceanic primary productivity. J. Plankton Res. **7**:691–702.

Melis, A. (1984) Light regulation of photosynthetic membrane structure, organization and function. J. Cell Biochim. **24**:271–285.

Melis, A. (1985) Functional properties of photosystem II in spinach chloroplasts. Biochim. Biophys. Acta **808**:334–342.

Melis, A., and Anderson, J.M. (1983) Structural and functional organization of the photosystems in spinach chloroplasts. Antenna size, relative electron-transport capacity and chlorophyll composition. Biochim. Biophys. Acta **724**:473–484.

Meunier, P.C., and Popovic, R. (1988) Optimization of the bare platinum electrode as an oxygen measurement system in photosynthesis. Photosynth. Res. **15**:271–279.

Mie, G. (1908) Beiträge zur optik trüber medien, speziell kolloidaler metallosungen. Ann. Phys. **25**:377–445.

Miles, C.J., and Brezonik, P.L. (1981) Oxygen consumption in humic-colored waters by a photochemical ferrous-ferric catalytic cycle. Environ. Sci. Tech. **15**:1089–1095.

Mishkind, M., Mauzerall, D., and Beale, S.I. (1979) Diurnal variation *in situ* of photosynthetic capacity in *Ulva* is caused by a dark reaction. Plant Physiol. **64**:896–899.

Mitchell, B.G., and Kiefer, D.A. (1984) Determination of absorption and fluorescence excitation spectra for phytoplankton. *In* Marine Phytoplankton and Productivity, O. Holm-Hansen, L. Bilis, and R. Gilles (Eds.) Springer-Verlag, Berlin, pp. 157–169.

Mitchell, B.G., and Kiefer, D.A. (1988a) Chlorophyll *a* specific absorption and fluorescence excitation spectra for light-limited phytoplankton. Deep-Sea Res. **35**:639–663.

Mitchell, B.G., and Kiefer, D.A. (1988b) Variability in pigment specific particulate fluorescence and absorption spectra in the northeastern Pacific Ocean. Deep-Sea Res. **35**:665–689.

Mohanty, P., and Govindjee (1973) Light-induced changes in the fluorescence yield of chlorophyll *a* in *Anacystis nidulans*. II. The fast changes and the effect of photosynthetic inhibitors on both the fast and slow fluorescent induction. Plant Cell Physiol. **14**:611–629.

Montgomery, H.A., Thom, N.S., and Cockburn, A. (1964) Determination of dissolved oxygen by the Winkler method, and the solubility of oxygen in pure water and in seawater. J. Appl. Chem. **14**:280–296.

Mopper, K., and Stahovec, W.L. (1986) Photochemical production of low molecular weight organic carbonyl compounds in seawater. Mar. Chem. **19**:305–321.

Moran, R., and Porath, D. (1980) Chlorophyll determination in intact tissues using N,N-dimethylformamide. Plant Physiol. **65**:478–479.

Morel, A., and Bricaud, A. (1981) Theoretical results concerning light absorption in a discrete medium, and applications to specific absorption of phytoplankton. Deep-Sea Res. **28A**:1375–1393.

Morel, A., and Bricaud, A. (1986) Inherent optical properties of algal cells including picoplankton: theoretical and experimental results. Can. Bull. Fish. Aquat. Sci. **214**:521–529.

Morel, A., and Prieur, L. (1977) Analysis of variations in ocean color. Limnol. Oceanogr. **22**:709–722.

Morel, A., and Smith, R.C. (1982) Terminology and units in optical oceanography. Mar. Geod. **5**:335–349.

Morel, N.M.L., Reuter, J.G., and Morel, R.M.M. (1978) Copper toxicity to *Skeletonema costatum* (Bacillariophyceae). J. Phycol. **14**:43–48.

Moreth, C.M. and Yentsch, C.S. (1970) A sensitive method for the determination of open ocean phytoplankton phycoerythrin pigments by fluorescence. Limnol. Oceanogr., **15**:313–317.

Morris, I., Yentsch, C.M., and Yentsch, C.S. (1971) Relationship between light and dark carbon dioxide fixation by marine algae. Limnol. Oceanogr. **16**:854–858.

Morris, I., Glover, H.E., and Yentsch, C.S. (1974) Products of photosynthesis by marine phytoplankton: the effect of environmental factors on the relative rates of protein synthesis. Mar. Biol. **27**:1–9.

Mortain-Bertrand, A., Descolos-Gros, C., and Jupin, H. (1987) Stimulating effect of light-to-dark transitions on carbon assimilation by a marine diatom. J. Exp. Mar. Biol. Ecol. **112**:11–26.

Mortain-Bertrand, A., Descolas-Gros, C., and Jupin, H. (1988) Pathway of dark inorganic carbon fixation in two species of diatoms: influence of light regime and regulator factors on diel variations. J. Plankton Res. **10**:199–217.

Murata, N., Nishimura, M., and Takamiya, A. (1966) Fluorescence of chlorophyll in photosynthetic systems. II. Induction of fluorescence in isolated spinach chloroplasts. Biochim. Biophys. Acta **120**:23–33.

Murphy, L.S., and Haugen, E.M. (1985) The distribution and abundance of phototrophic ultraplankton in the North Atlantic. Limnol. Oceanogr. **30**:47–58.

Murray, A.P., Gibbs, C.F., and Longmore, A.R. (1986) Determination of chlorophyll in marine waters: intercomparison of a rapid HPLC method with full HPLC, spectrophotometric and fluorometric methods. Mar. Chem. **19**:211–227.

Myers, J. (1970) Genetic and adaptive physiological characteristics observed in the chlorellas. *In* Prediction and Measurement of Photosynthetic Production. IBP/PP Technical Meeting, Trebon, 14–21 Sept., 1969. Wageningen, Pudoc pp. 447–454.

Myers, J. (1980) On the algae: thoughts about physiology and measurements of efficiency. *In* Primary Productivity in the Sea P.G. Falkowski (Ed.), Plenum Press, New York, pp. 1–15.

Myers, J., and Cramer, M. (1948) Metabolic conditions in *Chlorella*. J. Gen. Physiol. **32**:103–110.

Myers, J., and Graham, J.-R. (1971) The photosynthetic unit in *Chlorella* measured by repetitive short flashes. Plant Physiol. **48**:282–286.

Myers, J., and Graham, J.-R. (1983) On the ratio of photosynthetic reaction centers RC2/RC1 in *Chlorella*. Plant Physiol. **71**:440–442.

Myers, J., Graham, J.-R., and Wang, R.T. (1980) Light harvesting in *Anacystis nidulans* studied in pigment mutants. Plant Physiol. **66**:1144–1149.

Myklestad, S., Holm-Hansen, O., Varum, K.M., and Volcani, B.E. (1989) Rate of release of extracellular amino acids and carbohydrates from the marine diatom *Chaetoceros affinis*. J. Plankton Res. **11**:763–773.

Nalewajko, C., and Lean, D.R.S. (1972) Growth and excretion in planktonic algae and bacteria. J. Phycol. **8**:361–366.

Nalewajko, C., and Schindler, D.W. (1976) Primary production, extracellular release, and heterotrophy in two lakes in the ELA, Northwestern Ontario. J. Fish. Res. Bd. Can. **33**:219–226.

National Academy of Sciences (1984) Global Ocean Flux Study, Washington, DC.

Neale, P.J. (1987) Algal photoinhibition and photosynthesis in the aquatic environment. *In* Photoinhibition, D.J. Kyle, C.B. Osmond, and C.J. Arntzen, (Eds.), Elsevier, New York, pp. 39–65.

Neale, P.J., and Marra, J. (1985) Short-term variation of P_{max} under natural irradiance conditions: a model and its implications. Mar. Ecol. Prog. Ser. **26**:113–124.

Neale, P.J., and Melis, A. (1986) Algal photosynthetic membrane complexes and the photosynthesis-irradiance curve: a comparison of light-adaptation responses in *Chlamydomonas reinhardtii* (Chlorophyta). J. Phycol. **22**:531–538.

Nelson, S.G., and Siegrist, A.W. (1987) Comparison of mathematical expressions describing light-saturated photosynthesis by tropical marine macroalgae. Bull. Mar. Sci. **41**:617–622.

Neori, A., Vernet, M., Holm-Hansen, O., and Haxo, F.T. (1986) Relationship between action spectra for chlorophyll *a* fluorescence and photosynthetic O_2 evolution in algae. J. Plankton Res. **8**:357–348.

Neori, A., Vernet, M., Holm-Hansen, O., and Haxo, F.T. (1988) Comparison of chlorophyll far-red and red fluorescence excitation spectra with photosynthetic oxygen action spectra for photosystem II in algae. Mar. Ecol. Prog. Ser. **44**:297–302.

Neville, R.A., and Gower, J.F.R. (1977) Passive remote sensing of phytoplankton via chlorophyll *a* fluorescence. J. Geophys. Res. **82**:3487–3493.

Newport (1990) Precision laser and optics products catalogue. Newport Corporation, P.O. Box 8020, 18235 Mt. Baldy Circle, Fountain Valley, CA 92728-8029, U.S.A. or Pembroke House, Thompsons Close, Harpeldel, Herts AL54ES, U.K.

Oates, B.R., and Murray, S.N. (1983) Photosynthesis, dark respiration and desiccation resistance of the intertidal seaweeds *Hesperophycus harveyanus* and *Pelvetia fastigiata* F. *gracilis*. J. Phycol. **19**:371–380.

Odum, H.T. (1956) Primary production in flowing waters. Limnol. Oceanogr. **1**:102–117.

Odum, H.T., and Hoskin, C.M. (1958) Comparative studies on the metabolism of marine waters. Univ. Texas Inst. Mar. Sci. Contrib. **5**:16–46.

Odum, H.T., and Odum, E.P. (1955) Trophic structure and productivity of a windward coral reef community on Eniwetok Atoll. Ecol. Monogr. **25**:291–320.

Ogren, E., and Baker, N.R. (1985) Evaluation of a technique for the measurement of chlorophyll fluorescence from leaves exposed to continuous white light. Plant Cell Environ. **8**:539–547.

O'Leary, M.H. (1988) Carbon isotopes in photosynthesis. BioScience, **38**:325–336.

Oliver, R.L., and Ganf, G.G. (1988) The optical properties of a turbid reservoir and its phytoplankton in relation to photosynthesis and growth (Mount Bold Reservoir, South Australia). J. Plankton Res. **10**:1155–1177.

Olson, R.A., Brackett, F.S., and Crickard, R.G. (1949) Oxygen tension measurement by a method of time selection using the static platinum electrode with alternating potential. J. Gen. Physiol. **32**:681.

Ong, L.J., and Glazer, A.N. (1988) Structural studies of phycobiliproteins in unicellular marine cyanobacteria. *In* Light-Energy Transduction in Photosynthesis: Higher Plant and Bacterial Models, S.E. Stevens, Jr., and D.A. Bryant (Eds.), American Society of Plant Physiologists, Rockville, Md., pp. 102–121.

Ong, L.J., Glazer, A.N., and Waterbury, J.B. (1984) An unusual phycoerythrin from a marine cyanobacterium. Science **224**:80–83.

Öquist, G., Hallgren, J.-E., and Brunes, L. (1978) An apparatus for measuring photosynthetic quantum yields and quanta absorption spectra of intact plants. Plant Cell Environ. **1**:21–27.

Oriel Scientific (1990) Euro Catalogue Volume II. Light sources, monochromators and detection systems. 1 Mole Business Park, Leatherhead, Surrey, England or 250 Long Beach Boulevard, Stratford, Connecticut 06497 U.S.A.

Osborne, B.A., and R.J. Geider, (1986) Effects of nitrate limitation on photosynthesis in the diatom *Phaeodactylum tricornutum* Bohlin (Bacillariophyceae). Plant Cell Environ. **9**:617–625.

Osborne, B.A., and Geider, R.J. (1987a) The minimum photon requirement for photosynthesis. An analysis of the data of Warburg and Burk (1950) and Yuan, Evans and Daniels (1955). New Phytologist **106**:631–644.

Osborne, B.A., and Geider, R.J. (1987b) Photon requirement for growth of the diatom *Phaeodactylum tricornutum*. Plant Cell Environ. **10**:141–149.

Osborne, B.A., and Geider, R.J. (1989) Problems in the assessment of the package effect in five small phytoplankters. Mar. Biol. **100**:151–159.

Osborne, B.A., and Raven, J.A. (1986) Light absorption by plants and its implications for photosynthesis. Biol. Rev. **61**:1–61.

Otsuki, A., Ino, Y., and Fujii, T. (1983) Simultaneous measurements and determinations of stable carbon and nitrogen isotope ratios, and organic carbon and nitrogen contents in biological samples by coupling of a small quadruple mass spectrometer and modified carbon-nitrogen analyzer. Int. J. Mass Spectrom. Ion Phys. **48**:343–346.

Oudot, C. (1989) O_2 and CO_2 balances approach for estimating biological production in the mixed layer of the tropical Atlantic Ocean (Guinea Dome area). J. Mar. Res. **47**:385–409.

Oviatt, C.A., Rudnick, D.T., Keller, A.A., Sampou, P.A., and Almquist, T. (1986) A comparison of system (O_2 and CO_2) and ^{14}C measurements of metabolism in estuarine mesocosms. Mar. Ecol. Prog. Ser. **28**:57–67.

Paasche, E. (1963) The adaptation of the carbon-14 method for measurement of coccolith production in *Coccolithus huxlevi*. Physiol. Plant. **16**:186–200.

Paasche, E. (1964) A tracer study of inorganic carbon uptake during coccolith formation and photosynthesis in the cocclithophorid *Coccolithus huxlevi*. Plant Physiol. **35**(Suppl):1–82.

Palmisano, A.C., SooHoo, J.B., and Sullivan, C.W. (1987) Effects of four environmental variables on photosynthesis-irradiance relationships in Antartic sea-ice microalgae. Mar. Biol. **94**:299–306.

Pan, D., Gower, J.F.R., and Borstad, G.A. (1988) Seasonal variation of the surface chlorophyll distribution along the British Columbia coast as shown by CZCS satellite imagery. Limnol. Oceanog. **32**:227–244.

Papageorgiou, G. (1975) Fluorescence: an intrinsic probe of photosynthesis. *In* Govindjee, (Ed.), Bioenergetics of Photosynthesis, Academic Press, New York, pp. 320–371.

Park, K., Hood, D.W., and Odum, H.T. (1958) Diurnal pH variations in Texas Bays, and the application to primary production estimation. Univ. Texas Inst. Mar. Sci. Contrib. **5**:47–64.

Parsons, T.R., Maita, Y., and Lalli, C.M. (1984) A Manual of Chemical and Biological Methods for Seawater Analysis. Pergamon Press, New York.

Pearl, H.W. (1984) An evaluation of freeze fixation as a phytoplankton preservation method for microautoradiography. Limnol. Oceanogr. **29**:417–426.

Pearl, H.W., and Stull, E.A. (1979) In defense of grain density autoradiography. Limnol. Oceanogr. **23**:362–368.

Peeters, J.C.H., and Eilers, P. (1978) The relationship between light intensity and photosynthesis: a simple mathematical model. Hydrobiol. Bull. **12**:134–136.

Peltier, G., and Thibault, P. (1985) O_2 uptake in the light in *Chlamydomonas*. Plant Physiol. **79**:225–230.

Pennock, J.R., and Sharp, J.H. (1986) Phytoplankton production in the Delaware Estuary: temporal and spatial variability. Marine Ecol. Prog. Ser. **34**:143–155.

Pentacost, A. (1978) Calcification and photosynthesis in *Corallina officinalis* L. using $^{14}CO_2$ method. Brit. Phycol. J. **13**:383–390.

Pentecost, A. (1985) Photosynthetic plants as intermediary agents between environmental HCO_3^- and carbonate deposition. *In* Inorganic Carbon Uptake by Aquatic Photosynthetic Organisms, W.J. Lucas, J.A. Berry, (Eds.), Waverly Press, Baltimore, pp. 459–480.

Perry, M.J., Talbot, M.C., and Alberte, R.S. (1981) Photoadaption in marine phytoplankton: response of photosynthetic unit. Mar. Biol. **62**:91–101.

Petering, H.G., Duggar, B.M., and Daniels, F. (1939) Quantum efficiency of photosynthesis in *Chlorella*. J. Am. Chem. Soc. **61**:3525.

Peters, R.H. (1983) The Ecological Implications of Body Size, Cambridge University Press, Cambridge.

Petersen, R. (1982) Influence of copper and zinc on the growth of a freshwater alga, *Scenedesmusquadriculauda*: the significance of chemical speciation. Environ. Sci. Tech. **16**:443–447.

Peterson, B.J. (1980) Aquatic primary productivity and the ^{14}C-CO_2 method: a history of the productivity problem. Annu. Rev. Ecol. Syst. **11**:359–385.

Pfeifer, R.F., and McDiffett, W.F. (1975) Some factors affecting primary productivity of stream riffle communities. Arch. Hydrobiol. **75**:306–317.

Philipona, R. (1987) Simplified homodyne small-angle scattering technique which uses a diffraction pattern as local oscillator. Rev. Sci. Instrum. **58**:1572–??

Philips Lighting Comprehensive Handbook (1983) Philips Electronic and Associated Industires, Ltd., P.O. Box 298 City House, Droydon CR9 3QR, UK.

Phinney, D.A., and Yentsch, C.S. (1985) A novel phytoplankton chlorophyll technique: toward automated analysis. J. Plankton Res. **7**:633–642.

Pirt, S.J. (1986) The thermodynamic efficiency (quantum demand) and dynamics of photosynthetic growth. New Phytol. **102**:3–37.

Platt, T. (1984) Primary productivity in the central North Pacific: comparison of oxygen and carbon fluxes. Deep-Sea Res. **33**:1311–1319.

Platt, T. (1986) Primary production of the ocean water column as a function of surface light intensity: Algorithms for remote sensing. Deep-Sea Res. **33**:149–163.

Platt, T., and Harrison, W.G. (1986a) Biogenic fluxes of carbon and oxygen in the ocean. Nature **318**:55–58.

Platt, T., and Harrison, W.G. (1986b) Reconciliation of carbon and oxygen fluxes in the upper ocean. Deep-Sea Res. **33**:273–276.

Platt, T., and Herman, A.W. (1983) Remote sensing of phytoplankton in the sea: surface-layer chlorophyll as an estimate of water-column chlorophyll and primary production. Int. J. Remote Sensing **4**:343–351.

Platt, T., Gallegos, C.L., and Harrison, W.G. (1980). Photoinhibition of photosynthesis in natural assemblages of marine phytoplankton. J. Mar. Res. **38**:687–701.

Platt, T., Lewis, M., and Geider, R. (1984) Thermodynamics of the pelagic ecosystem: elementary closure conditions for biological production in the open ocean. *In* Flows of Energy and Materials in Marine Ecosystems, M.J.R. Fasham (Ed.), Plenum Publishing, New York, pp. 49–84.

Platt, T., Harrison, W.G., Horne, E.P., and Irwin, B. (1987) Carbon fixation and oxygen evolution by phytoplankton in the Canadian High Arctic. Polar Biol. **8**:103–114.

Platt, T., Sathyendranath, S., Caverhill, C.M., and Lewis, M.R. (1988) Ocean primary production and available light: further algorithms for remote sensing. Deep-Sea Res. **35**:855–879.

Pokorny, J., Orr, P.T., Ondok, J.P., and Denny, P. (1989) Photosynthetic quotients of some aquatic macrophyte species. Photosynthetica **23**:494–506.

Porra, R.J., Thompson, W.A., and Kriedemann, D.E. (1989) Determination of accurate extinction coefficients and simultaneous equations for assaying chlorophylls *a* and *b* extracted with four different solvents: verification of the concentration of chlorophyll standards by atomic absorption spectroscopy. Biochim. Biophys. Acta **975**:384–394.

Pregnall, A.M. (1983) Release of dissolved organic carbon from the estuarine intertidal macroalga *Enteromorpha prolifera*. Mar. Biol. **73**:37–42.

Preisendorfer, R.W. (1961) Application of radiative transfer theory to light measurements in the sea. Union Geod. Geophys. Inst. Monogr. **10**:11–30.

Preston, T., and Owens, N.J.P. (1985) Preliminary ^{13}C measurements using a gas chromatograph interfaced to an isotope ratio mass spectrometer. Biomed. Mass Spectrom. **12**:510–513.

Prézelin, B.B. (1976) The role of peridinin-chlorophyll *a*-proteins in the photosynthetic light adaption of the marine dinoflagellate, *Glenodinium* sp. Planta **130**:225–233.

Prézelin, B.B., and Sweeney, B.M. (1978) Photoadaptation of photosynthesis in *Gonyaulax polyedra*. Mar. Biol. **48**:27–35.

Prézelin, B.B., and Sweeney, B.M. (1977) Characterization of photosynthetic rhythms in marine dinoflagellates. II. Photosynthesis-irradiance curves and *in vivo* chlorophyll *a* fluorescence. Plant Physiol. **60**:388–392.

Price, N.M., Harrison, P.J., Landry, M.R., Azam, F., and Hall, K.J.F. (1986) Toxic effects of latex, and tygon tubing on marine phytoplankton, zooplankton and bacteria. Mar. Ecol. Prog. Ser. **34**:41–49.

Priscu, J.C., and Goldman, C.R. (1984) The effect of temperature on photosynthetic and respiratory electron transport system activity in the shallow and deep-living phytoplankton of a subalpine lake. Freshwater Biol. **14**:143–155.

Privoznik, K.G., Daniel, K.J., and Incropera, F.P. (1978) Absorption, extinction and phase function measurements for algal suspensions of *Chlorella pyrenoidosa*. J. Quant. Spectrosc. Radiat. Trans. **20**:345–352.

Quadir, A., Harrison, P.J., and DeWreede, R.E. (1979) The effects of emergence and submergence on the photosynthesis and respiration of marine macrophytes. Phycologia **18**:83–88.

Rabideau, G.S., French, C.S., and Holt, A.S. (1946) The absorption and reflection spectra of leaves, chloroplast suspension, and chloroplast fragments as measured in an Ulbricht sphere. Am. J. Bot. **33**:769–777.

Radmer, R., and Ollinger, O. (1980) Light-driven uptake of oxygen, carbon dioxide and bicarbonate in the green alga *Scenedesmus*. Plant Physiol. **65**:723–729.

Raine, R.C.T. (1983) The effect of nitrogen supply on the photosynthetic quotient of natural phytoplankton assemblages. Bot. Mar. **26**:417–423.

Ramus, J. (1978) Seaweed anatomy and photosynthetic performance: the ecological significance of light guides, heterogeneous absorption and multiple scatter. J. Phycol. **14**:352–362.

Ramus, J. (1985) Light. *In* Handbook of Phycological Methods. Ecological Field Methods: Macroalgae, M.M. Littler and D.S. Littler (Eds.), Cambridge University Press, New York, pp. 33–52.

Ramus, J., and Rosenberg, G. (1980) Diurnal photosynthetic performance of seaweeds under natural conditions. Mar. Biol. **56**:21–28.

Raps, S., Wyman, K., Siegelman, H.W., and Falkowski, P.G. (1983) Adaptation of the cyanobacterium *Microcystis aeruginosa* to light intensity. Plant Physiol. **72**:829–832.

Raven, J.A. (1984) Energetics and Transport in Aquatic Plants. A.R. Liss, New York.

Raven, J.A., and Geider, R.J. (1988) Temperature and algal growth. New Phytol. **110**:441–461.

Redalje, D.G. (1983) Phytoplankton carbon biomass and specific growth rates determined with the labeled chlorophyll *a* technique. Mar. Ecol. Prog. Ser. **11**:217–225.

Redalje, D.G., and Laws, E.A. (1981) A new method for estimating phytoplankton growth rates and carbon biomass. Mar. Biol. **62**:73–79.

Redfield, A.C. (1948) The exchange of oxygen across the sea surface. J. Mar. Res. **7**:347–361.

Renger, G. and Schreiber, U. (1986) Practical applications of fluorometric methods to algae and higher plant research. *In* Light Emission by Plants and Bacteria, Govindjee, J. Amesz, and D.C. Fork (Eds.), Academic Press, Orlando, pp. 587–638.

Reule, Von A. (1962) Zur theorie der Taylor-Kugel. Zeiss-Mitteilungen **2**:371–387.

Revsbech, N.P., and Jorgensen, B.B. (1986) Microelectrodes: their use in microbial ecology. Adv. Microbiol. Ecol. **9**:293–352.

Riaux-Gobin, C., Llewellyn, C.A., and Klein, B. (1987) Microphytobenthos from two subtidal sediments from North Brittany. II. Variations of pigment compositions and concentrations determined by HPLC and conventional techniques. Mar. Ecol. Prog. Ser. **40**:275–283.

Richards, L., and Thurston, C.F. (1980) Protein turnover in *Chlorella vulgaris* var. *vacuolata*: measurement of overall rate of intracellular protein degradation using isotope exchange with water. J. Gen. Microbiol. **121**:49–61.

Richardson, K., Beardall, J., and Raven, J.A. (1983) Adaptation of unicellular algae to irradiance: an analysis of strategies. New Phytol. **93**:157–191.

Riley, G.A. (1938) Plankton studies. I. A preliminary investigation of the plankton of the Tortugas region. J. Mar. Res. **1**:335–352.

Riley, G.A. (1939) Plankton studies. II. The western North Atlantic, May–June, 1939. J. Mar. Res., **2**:145–162.

Riley, G.A. (1941) Plankton studies. V. Regional summary. J. Mar. Res. **4**:162–171.

Riley, G.A. (1944) The carbon metabolism and photosynthetic efficiency of the earth as a whole. Am. Sci. **32**:129–134.

Riper, D.M., Owens, T.G. and Falkowski, P.G. (1979) Chlorophyll turnover in *Skeletonema costatum*, a marine plankton diatom. Plant Physiol., **64**:49–54.

Rivas, J. de las, Abadia, A., Abadia, J. (1989) A new reversed phase-HPLC method resolving all major higher plant photosynthetic pigments. Plant Physiol. **91**:190–192.

Rivkin, R.B. (1985) Carbon-14 labelling patterns of individual marine phytoplankton from natural populations: a species approach. Mar. Biol. **89**:135–142.

Rivkin, R.B., and Putt, M. (1987) Heterotrophy and photoheterotrophy by Antarctic microalgae: light-dependent incorporation of amino acids and glucose. J. Phycol. **23**:442–452.

Rivkin, R.B., and Seliger, H.H. (1981) Liquid scintillation counting for ^{14}C uptake of single algal cells isolated from natural samples. Limnol. Oceanogr. **26**:780–785.

Rivkin, R.B., and Voytek, M.A. (1986) Cell division rates of eucaryotic algae measured by tritiated thymidine incorporation into DNA: coincident measurements of photosynthesis and cell division of individual species of phytoplankton isolated from natural populations. J. Phycol. **22**:199–205.

Rivkin, R.B., Seliger, H.H., Swift, E., and Biggley, W.H. (1982) Light-shade adaptation by the oceanic dinoflagellates *Pyrocystis noctiluca* and *P. fusiformis.* Mar. Biol. **68**:181–191.

Rivkin, R.B., Swift, E., Seiger, H.H. and Biggley, W.H., and Voytek, M.A. (1984) Growth and carbon uptake by natural populations of *Pyrocystis noctiluca* and *P. fusiformis.* Deep-Sea Res. **31**:353–364.

Roberts, D.G., and Smith, D.M. (1988) Infrared gas analysis of both gaseous and dissolved CO_2 in small-volume marine samples. Limnol. Oceanogr. **33**:135–140.

Roberts, R.B., Abelson, P.H., Cowie, D.B., Botton, E.T., and Britten, R.J. (1963) Studies of biosynthesis in *Escherichia coli.* Carnegie Inst. Wash, Publ. 607, 521 pp.

Romero, J.M., Lara, C., and Sivak, M.N. (1989) Changes in net O_2 exchange induced by inorganic nitrogen in the blue-green alga *Anacystis nidulans.* Plant Physiol. **91**:28–30.

Ruhle, W., and Wild, A. (1979) The intensification of absorbance changes in leaves by light-dispersion. Differences between high-light and low-light leaves. Planta **146**:551–557.

Rurainski, H.J., and Mader, G. (1980) Light-emitting diodes as a light source in photochemical and photobiological work. Meth. Enzymol. **69**:667–675.

Ryther, J.H. (1956) The measurement of primary production. Limnol. Oceanogr. **1**:72–84.

Ryther, J.H. (1969) Photosynthesis and fish production in the sea. Science **166**:72–76.

Ryther, J.H., and Yentsch, C.S. (1957) The estimation of phytoplankton production in the ocean from chlorophyll and light data. Limnol. Oceanogr. **2**:281–286.

Safwat, H.H. (1970) Effect of centrally located samples in the integrating sphere. J. Opt. Soc. Am. **60**:534–541.

Sakamoto, M. Tilzer, M.M., Gächter, R., Rai, H., Collos, Y., Tschumi, P., Berner, P., Zbaren, D., Zbaren, J., Dokulil, M., Bossard, P., Uehlinger, U., and Nusch, E.A. (1984) Joint field experiments for comparisons of measuring methods of photosynthetic production. J. Plankton Res. **6**:365–383.

Sakshaug, E., Andresen, K., and Kiefer, D.A. (1989) A steady state description of growth and light absorption in the marine diatom *Skeletonema costatum.* Limnol. Oceanogr. **34**:198–205.

Salonen, K. (1979) Comparison of different glass fibre and silver metal filters for the determination of particulate organic carbon. Hydrobiol. **67**:29–32.

Sane, P.V., Tatake, V.G., and Desai, T.S. (1974) Detection of the triplet states of chlorophylls *in vivo.* FEBS Lett. **45**:290–294.

Sarmiento, J.L., and Toggweiler, J.R. (1984) A new model for the role of the ocean in determining atmospheric pCO_2. Nature **308**:621–624.

Sathyendranath, S. (1986) Remote sensing of phytoplankton: a review, with special reference to picoplankton. Can. Bull. Fish. Aquat. Sci. **214**:561–583.

Sathyendranath, S., and Platt, T. (1989) Computation of aquatic primary production: extended formalism to include effect of angular and spectral distribution of light. Limnol. Oceanogr. **34**:188–198.

Sathyendranath, S., Lazzara, L., and Prieur, L. (1987) Variations in the spectral values of specific absorption of phytoplankton. Limnol. Oceanogr. **32**:403–415.

Schafer, C., and Bjorkman, V. (1989) Relationship between efficiency of photosynthetic energy conversion and chlorophyll fluorescence quenching in upland cotton (*Gossypium hirsutum* L.). Planta **178**:367–376.

Schindler, D.W. (1966) A liquid scintillation method for measuring carbon-14 uptake in photosynthesis. Nature **211**:844–845.

Schindler, D.W., Schmidt, R.V., and Reid, D.A. (1972) Acidification and bubbling as an alternative to filtration in determining phytoplankton production by the ^{14}C method. J. Fish. Res. Bd. Can. **29**:1627–1631.

Schlesinger, D.A., Molot, L.A., and Shuter, B.J. (1981) Specific growth rates of freshwater algae in relation to cell size and light intensity. Can. J. Fish. Aquat. Sci. **9**:1052–1058.

Schlesinger, D.A., and Shuter, B.J. (1981) Patterns of growth and cell composition of freshwater algae in light-limited continuous cultures. J. Phycol. **17**:250–256.

Scholander, P.F., Dam, L. van, Claff, C.L., and Kanwisher, J.W. (1955) Micro gasometric determination of dissolved oxygen and nitrogen. Biol. Bull. **109**:328–334.

Schreiber, U. (1983) Chlorophyll fluorescence yield changes as a tool in plant physiology. I. The measuring system. Photosynth. Res. **4**:361–373.

Schreiber, U. (1986) Detection of rapid induction kinetics with a new type of high-frequency modulated chlorophyll fluorometer. Photosynth. Res. **9**:261–272.

Schreiber, U., and Bilger, W. (1987) Rapid assessment of stress effects on plant leaves by chlorophyll fluorescence measurements. *In* Plant Response to Stress—Functional Analysis in Mediterranean Ecosystems J.D. Tenhunen, R.M. Catanino, O.L. Lange, W.C. Vechel (Eds.), Springer-Verlag, Berlin, pp. 27–53.

Schreiber, U., and Schliwa, U. (1987) A solid-state, portable instrument for measurement of chlorophyll luminescence induction in plants. Photosynth. Res. **11**:173–182.

Schreiber, U., Schliwa, U., and Bilger, W. (1986) Continuous recording of photochemical and non-photochemical fluorescence quenching with a new type of modulation fluorometer. Photosynth. Res. **10**:51–62.

Schreiber, U., Neubauer, C., and Klughammer, C. (1989) Devices and methods for room temperature fluorescence analysis. Phil. Trans. R. Soc. Lond. **B323**:241–251.

Schwartz, S.J., and von Elbe, J.H. (1982) High performance liquid chromatography of plant pigments—a review. J. Liquid Chrom. **5**(1):43–73.

Scott, B.D., and Jitts, H.R. (1977) Photosynthesis of phytoplankton and zooxanthellae on a coral reef. Mar. Biol. **41**:307–315.

Senft, W.H. (1978) Dependence of light-saturated rates of algal photosynthesis on intracellular concentrations of phosphorus. Limnol. Oceanogr. **23**:709–718.

Seyfried, M., and Fukshansky, L. (1983) Light gradients in plant tissue. Applied Optics, 22:1402–1408.

Sharkey, T.D. (1989) Evaluationg the role of Rubisco regulation in photosynthesis of C_3 plants. Phil. Trans. R. Soc. Lond. B323:435–448.

Sharkey, T.D., and Berry, J.A. (1985) Carbon isotope fractionation of algae as influenced by an inducible CO_2 concentrating mechanism. In Inorganic Carbon Uptake by Aquatic Organisms. W.J. Lucas and J.A. Berry (Eds.), Am. Soc. Plant Physiol., Rockville, MD, pp. 389–401.

Sharkey, T.D., Seeman, J.R., and Pearcy, R.W. (1986a) Contribution of metabolites of photosynthesis to postillumination CO_2 assimilation in response to light flecks. Plant Physiol. 82:1063–1068.

Sharkey, T.D., Stitt, M., Heineke, D., Gerhardt, R., Rasche, K., and Heldt, H.W. (1986b) Limitation of photosynthesis by carbon metabolism. II. O_2 insensitive CO_2 uptake results from limitation of triose phosphate utilization. Plant Physiol. 81:1123–1129.

Sharp, J.H. (1977) Excretion of organic matter by marine phytoplankton: do healthy cells do it? Limnol. Oceanogr. 22:381–399.

Sharp, R.E., Matthews, M.A. and Boyer J.S. (1984) Kok effect and the quantum yield of photosynthesis. Plant Physiol., 75, 95–101.

Sheldon, R.W. (1972) Size separation of marine seston by membrane and glass-giber filters. Limnol. Oceanogr. 17:494–498.

Shibata, K., Benson, A.A., and Calvin, M. (1954) The absorption spectra of suspensions of living microorganisms. Biochim. Biophys. Acta 15:461–470.

Shifrin, N.S., and Chisholm, S.W. (1981) Phytoplankton lipids: interspecific differences and effects of nitrate, silicate and light-dark cycles. J. Phycol. 17:374–384.

Shoaf, W.T., and Lium, B.W. (1976) Improved extraction of chlorophyll a and b from algae using dimethyl sulfoxide. Limnol. Oceanogr. 21:926–928.

Shulenberger, E., and Reid, J.L. (1981) The Pacific shallow oxygen maximum, deep chlorophyll maximum, and primary productivity reconsidered. Deep-Sea Res. 28:901–919.

Shuter, B. (1979) A model of physiological adaptation in unicellular algae. J. Theor. Biol. 78:519–552.

Siegelman, H.W., and Kycia, J.H. (1978) Algal biliproteins. In Handbook of Phycological Methods. Physiological and Biochemical Methods, J.A. Hellebust and J.S. Craigie (Eds.), Cambridge University Press, Cambridge, pp. 71–79.

Silver, M.W., and Davoll, P.J. (1978) Loss of ^{14}C activity after chemical fixation of phytoplankton: error source for autoradiography and other productivity measurements. Limnol. Oceanogr. 2:362–368.

Skirrow, G. (1975) The dissolved gases—carbon dioxide. In Chemical Oceanography, Vol. 2, J.P. Riley and G. Skirrow (Eds.), Academic Press, New York, pp. 1–192.

Slawyk, G., Minas, M., Collos, Y., Legendre, L., and Roy, S. (1984) Comparison of radioactive and stable isotope tracer techniques for measuring photosynthesis: ^{13}C and ^{14}C uptake by marine phytoplankton. J. Plankton Res. 6:249–257.

Slawyk, G., L'Helguen, S., Collos, Y., and Freije, H. (1988) Quantitative determination of particulate organic N and C in marine-phytoplankton samples using mass-spectrometer signals from isotope-ratio analyses in ^{15}N- and ^{13}C-tracer studies. J. Exp. Mar. Biol. Ecol. **115**:187–195.

Smith, D.F. (1982) Observation and quantitative analysis of curvilinear regions of time-varying oxygen concentration with an oxygen electrode and a minicomputer. J. Exp. Mar. Biol. Ecol. **64**:117–124.

Smith, D.F., and Horner, S.M.J. (1981) Tracer kinetic analysis applied to problems in marine biology. Can. Bull. Fish. Aquat. Sci. **210**:113–129.

Smith, F.A., and Raven, J.A. (1976) H$^+$ transport and regulation of cell pH. *In* Encyclopedia of Plant Physiology, U. Lüttge, and M.G. Pitman (Eds.), Springer, Berlin, pp. 317–346.

Smith, F.A., and Walker, N.A. (1980) Photosynthesis by aquatic plants: effects of unstirred layers in relation to assimilation of CO_2 and HCO_3 and to carbon isotopic discrimination. New Phytol. **86**:245–259.

Smith, H., and Holmes, M.G. (1984) Techniques in Photomorphogenesis. Academic Press, London.

Smith, R.C. (1981) Remote sensing and depth distribution of ocean chlorophyll. Mar. Ecol. Prog. Ser. **5**:359–361.

Smith, R.C., Eppley, R.W., and Baker, K.S. (1982) Correlation of primary production as measured aboard ship in southern California coastal waters and as estimated from satellite chlorophyll images. Mar. Biol. **66**:281–288.

Smith, R.E.H., and Geider, R.J. (1985) Kinetics of intracellular carbon allocation in a marine diatom. J. Exp. Mar. Biol. Ecol. **93**:191–210.

Smith, R.E.H., and Platt, T. (1984) Carbon exchange and ^{14}C tracer methods in a nitrogen-limited diatom, *Thalassiosira pseudonana*. Mar. Ecol. Prog. Ser. **16**:75–87.

Smith, R.E.H., Geider, R.J., and Platt, T. (1984) Microplankton productivity in the oligotrophic ocean. Nature **311**:252–254.

Smith, S.V., and Key, G.S. (1975) Carbon dioxide and metabolism in marine environment. Limnol. Oceanogr. **20**:493–495.

Smith, S.V., and Kinsey, D.W. (1978) Calcification and organic carbon metabolism as indicated by carbon dioxide. *In* Coral Reefs: Research Methods, D.R. Stoddart and R.E. Johannes, (Eds.), UNGSCO, New York, pp. 469–484.

Smith, V.H. (1979) Nutrient dependence of primary productivity in lakes. Limnol. Oceanogr. **24**:1051–1064.

Smith, V.H. (1983) Light and nutrient dependence of photosynthesis by algae. J. Phycol. **19**:306–313.

Smucker, R.A., and Dauso, R. (1986) Products of photosynthesis by marine phytoplankton: chitin in TCA "protein" precipitates. J. Exp. Mar. Biol. Ecol. **104**:143–152.

Sondergaard, D.F. and Schieup, H. (1982) Release of extracellular organic carbon during a diatom bloom in Lake Mosso: molecular weight fractionation. Freshwater Biol., **12**:313–320.

SooHoo, J.B., Kiefer, D.A., Collins, D.J., and McDermid, I.S. (1986) *In vivo* fluorescence excitation and absorption spectra of marine phytoplankton: I.

Taxonomic characteristics and responses to photoadaptation. J. Plankton Res. 8:197–214.

Sournia, A. (1974) Circadian periodicities in natural populations of marine phytoplankton. Adv. Mar. Biol. 12:325–389.

Spinrad, R.W., Zaneveld, J.R.V., and Pak, H. (1978) Volume scattering function of suspended particulate matter at near-forward angles: a comparison of experimental and theoretical values. Appl. Opt. 17:1125–1130.

Spitzer, W.S., and Jenkins, W.J. (1989) Rates of vertical mixing, gas exchange and new production: estimates from seasonal gas cycles in the upper ocean near Bermuda. J. Mar. Res. 47:169–196.

Stauffer, R.E., Lee, G.F., and Armstrong, D.E. (1979) Estimating chlorophyll extraction biases. J. Fish. Res. Board Can. 36:152–157.

Steeman-Nielsen, E. (1952) The use of radio-active carbon (^{14}C) for measuring organic production in the sea. J. Cons. Int. Explor. Mer. 18:117–140.

Steeman-Nielsen, E., and Bruum-Laursen, H. (1976) Effect of $CuSO_4$ on the photosynthetic rate of phytoplankton in four Danish lakes. Oikos 27:293–242.

Steinberg, C. (1978) Freisetzung gelosten organischen Kohlenstoffs (DOC) verschiedener Molekugrossen in Planktongesellschaften. Arch. Hydrobiol. 82:155–165.

Stephenson, R.L., Tan, F.C., and Mann, K. (1984) Stable carbon isotope variability in marine macrophytes and its implications for food web studies. Mar. Biol. 81:223–230.

Stewart, D.E., and Farmer, F.H. (1984) Extraction, identification, and quantification of phycobiliprotein pigments from phototrophic plankton. Limnol. Oceanogr. 29:392–397.

Stitt, M. (1986) Limitation of photosynthesis by carbon metabolism. I. Evidence for excess electron transport capacity in leaves carrying out photosynthesis in saturating light and CO_2. Plant Physiol. 81:1115–1122.

Stockner, J.D. (1988) Phototrophic picoplankton: An overview from marine and freshwater ecosystems. Limnol. Oceanogr. 33:765–775.

Strickland, J.D.H. (1960) Measuring the production of marine phytoplankton. Bull. Fish. Res. Bd. Can. 122:1–172.

Strickland, J.D.H., and Parsons, T.R. (1972) A practical handbook of seawater analysis. Bull. Fish. Res. Bd. Can. 167:1–310.

Stull, E.A., de Amezaga, E., and Goldman, C.R. (1973) The contribution of individual species of algae to primary productivity of Castle Lake, California. Verh, Int. Verein. Limnol. 18:1776–1783.

Stumm, W., and Morgan, J.J. (1970) Aquatic Chemistry, Wiley-Interscience, New York.

Stumpf, R.P., and Tyler, M.A. (1988) Satellite detection of bloom and pigment distributions in estuaries. Remote Sensing Environ. 24:385–404.

Süeltemeyer, D.F., Klug, K., and Fock, H.P. (1986) Effect of photon fluence rate on oxygen evolution and uptake by Chlamydomonas reinhardtii suspensions grown in ambient and CO_2-enriched air. Plant Physiol. 81:372–375.

Süeltemeyer, D.F., Klug, K. and Fock, H.P. (1986) Effect of photon fluence rate on oxygen evolution and uptake by *Chlamydomonas reinhardtii* suspensions grown in ambient and CO_2-enriched air. Plant Physiol., **81**:372–375.

Sukenik, A., Bennet, J., and Falkowski, P. (1987) Light-saturated photosynthesis—limitation by electron transport or carbon fixation? Biochim. Biophys. Acta **891**:205–215.

Sunda, W.G., and Guillard, R.R. (1976) The relationship between cupric ion activity and the toxicity of copper to phytoplankton. J. Mar. Res. **34**:S11–S29.

Sundh, I. (1989) Characterization of phytoplankton extracellular products (PDOC) and their subsequent uptake by heterotrophic organisms in a mesotrophic forest lake. J. Plankton Res. **11**:463–486.

Suzuki, R., and Fujita, Y. (1986) Chlorophyll decomposition in *Skeletonema costatum*: a problem in chlorophyll determination of water samples. Mar. Ecol. Prog. Ser. **28**:81–85.

Tabor, P.S., and Neihof, R.A. (1982) Improved microautoradiographic method to determine individual microorganisms active in substrate uptake in natural waters. Appl. Environ. Microbiol. **44**:945–953.

Taguchi, S. (1976) Relationship between photosynthesis and cell size of marine diatoms. J. Physiol. **12**:185–189.

Taguchi, S. (1983) Dark fixation of CO_2 in the subtropical North Pacific Ocean and the Weddell Sea. Bull. Plankton Soc. Jap. **30**:115–124.

Taguchi, S., and Laws, E.A. (1985) Application of a single-cell isolation technique to studies of carbon assimilation by the subtropical silicoflagellate *Dictyocha perlaevis*. Mar. Ecol. Prog. Ser. **23**:251–255.

Taguchi, S., and Laws, E.A. (1988) On the microparticles which pass through glass fiber type GF/F in coastal and open waters. J. Plankton Res. **10**:999–1008.

Talling, J.F. (1957a) Photosynthetic characteristics of some freshwater plankton diatoms in relation to underwater radiation. New Phytol. **56**:29–50.

Talling, J.F. (1967b) Diurnal changes of stratification and photosynthesis in some tropical African waters. Proc. Roy. Soc. Lond. **B147**:57–83.

Talling, J.F. (1965) The photosynthetic activity of phytoplankton in East African lakes. Int Rev. Ges. Hydrobiol. **50**:1–32.

Talling, J.F. (1976) The depletion of carbon dioxide from lake water by phytoplankton. J. Ecol. **64**:79–121.

Talling, J.F. (1984) Past and contemporary trends and attitudes in work on primary productivity. J. Plankton Res. **6**:203–217.

Taylor, A.H. (1920) The measurement of diffuse reflection factors and a new absolute reflectometer. J. Opt. Soc. Am. **4**:9–23.

Terashima, I. (1986) Dorsiventrality in photosynthetic light response curves of a leaf. J. Exp. Bot. **37**:399–405.

Terashima, I., and Inoue, Y. (1985a) Palisade tissue chloroplasts and spongy tissue chloroplasts in spinach: Biochemical and ultrastructural differences. Plant Cell Physiol. **26**:63–75.

Terashima, I., and Inoue, Y. (1985b) Vertical gradient in photosynthetic properties of spinach chloroplasts dependent on intra-leaf light environment. Plant Cell Physiol. **26**:781–785.

Terashima, I., and Saeki, T. (1983) Light environment within a leaf. I. Optical properties of paradermal sections of *Camellia* leaves with special reference to differences in the optical properties of palisade and spongy tissues. Plant Cell Physiol. **24**:1493–1501.

Terry, K.L. (1982) Nitrate uptake and assimilation in *Thalassiosira weissflogii* and *Phaeodactylum tricornutum*: interactions with photosynthesis and with the uptake of other ions. Mar. Biol. **69**:21–30.

Theodórsson, I.T., and Bjarnason, J.O. (1975) The acid-bubbling method for primary productivity measurements modified and tested. Limnol. Oceanogr. **20**:1018–1019.

Thielen, A.P.G.M., and Van Gorkom, H.J. (1981) Redox potentials of electron acceptors in photosystem IIα and IIβ. Biochim. Biophys. Acta **129**:205–209.

Thomas, W.H. (1964) An experimental evaluation of the ^{14}C method for measuring phytoplankton production, using cultures of *Dunaliella tertiolecta*. Fishery Bull. **63**:273–292.

Thornley, J.H.M. (1976) Mathematical Models in Plant Physiology, Academic Press, London.

Thorn Lighting Technical Handbook (1989) Tate (Ed.), Thorn Lighting.

Tijssen, S.B. (1979) Diurnal oxygen rhythm and primary production in the mixed layer of the Atlantic Ocean at 20°N. Neth. J. Sea Res. **13**:79–84.

Tilzer, M.M., Elbrachter, M., Gieskes, W.W., and Beese, B. (1986) Light-temperature interactions in the control of photosynthesis in Antarctic Phytoplankton. Polar Biol. **5**:105–111.

Topliss, B.J., and Platt, T. (1986) Passive fluorescence and photosynthesis in the ocean: implications for remote sensing. Deep-Sea Res. **33**:849–864.

Trebst, A.V., Tsujimoto, H.Y., and Arnon, D.I. (1958) Separation of light and dark phases in the photosynthesis of isolated chloroplasts. Nature **182**:351–355.

Trees, C.C., Kennicutt, M.C., and Brooks, J.M. (1985) Errors associated with the standard fluorimetric determinations of chlorophylls and phaeopigments. Mar. Chem. **17**:1–12.

Trees, C.C., Bidigare, R.R., and Brooks, J.M. (1986) Distribution of chlorophylls and phaeopigments in the Northwestern Atlantic Ocean. J. Plank. Res. **8**:447–458.

Tregunna, E.B., and Thomas, E.A. (1968) Measurement of inorganic carbon and photosynthesis in sea water by pCO_2 and pH analysis. Can J. Bot. **46**:481–485.

Tsuji, T., Ohki, K., and Fujita, Y. (1986) Determination of photosynthetic pigment composition in an individual phytoplankton cell in seas and lakes using fluorescence microscopy; properties of the fluorescence emitted from picophytoplankton cells. Mar. Biol. **93**:343–349.

Ulbricht, R. (1920) Das Kugelphotometer (Ulbricht'sche Kugel), Oldenbourg, Munich.

UNESCO (United Nations Educational, Scientific and Cultural Organization), (1973) A Guide to the Measurement of Marine Primary Production Under Some Special Conditions, UNESCO, Paris.

Vanderkooi, J.M., and Berger, J.W. (1989) Excited triplet states used to study biological macromolecules at room temperature. Biochim. Biophys. Acta **976**:1–27.

Vesk, M., and Jeffrey, S.W. (1987) Ultrastructure and pigments of two strains of the picoplanktonic alga *Pelagococcus subviridis* (Chrysophyceae). J. Phycol. **23**:322–336.

Vinberg, G.G. (1940) Measurement of the rate of exchange of oxygen between a water basin and the atmosphere. C.R. Acad. Sci. USSR **26**:666–669.

Vinberg, G.G., and Yarovitzina, L.I. (1939) Daily changes in the quantity of dissolved oxygen as a method for measuring the value of primary production. (Russ., Eng. summary.) Arb. Limnol. Sta. Kossino **22**:128–143.

Vogelmann, T.C., and Björn, L.O. (1984) Measurement of light gradients and spectral regime in plant tissue with a fiber optic probe. Physiol. Plant **60**:361–368.

Vogelmann, T.C., Knapp, A., McClean, T., and Smith, W.K. (1988) Measurement of light gradients within thin plant tissues with fiber optic microsensors. Physiol. Plant **72**:623–630.

Vollenweider, R.A. (1969) A Manual on Methods for Measuring Primary Production in Aquatic Environments, IBP Handbook No. 12, Blackwell Scientific Publishers, Oxford.

Vonshak, A., Sivack, M., and Walker, D. (1989) Use of a solid support in the study of photosynthetic activity of the cyanobacterium *Spirulina platensis*. J. Appl. Phycol. **1**:131–135.

Walker, D.A. (1981) Photosynthetic induction. *In* Proceedings of the 5th International Congress on Photosynthesis, Halkidiki, Greece, G. Akoyonoglou (Ed.), Vol. 4, Balaban, Philadelphia, pp. 189–202.

Walker, D.A. (1989) Automated measurement of leaf photosynthetic O_2 evolution as a function of photon flux density. Phil. Trans. R. Soc. Lond. **B323**:313–326.

Walker, D.A., and Osmond, C.B. (1986) Measurements of photosynthesis *in vivo* with a leaf disc electrode: correlations between light-dependence of steady-state photosynthetic O_2 evolution and chlorophyll *a* fluorescence transients. Proc. R. Soc. Lond. **B227**:267–280.

Walker, D.A., Horton, P., Sivak, M., and Quick, W.P. (1983) Antiparallel relationships between O_2 evolution and slow fluorescence induction kinetics. Photobiochem. and Photobiophys. **5**:35–39.

Wanner, U., and Egli, T. (1990) Dynamics of microbial growth and cell composition in batch culture. FEMS Microbiol. Rev. **75**:19–44.

Wantanabe, Y. (1980) A study of the excretion and extracellular products of natural phytoplankton in Lake Nakanuma, Japan. Int. Rev. Gas. Hydrobiol. **65**:809–834.

Watt, W.D. (1966) Release of dissolved organic material from the cells of phytoplankton populations. Proc. R. Soc. Lond. **B164**:521–551.

Watt, W.D. (1971) Measuring the primary production rates of individual phytoplankton species in natural mixed populations. Deep-Sea Res. **18**:329–339.

Weger, H.G., and Turpin, D.H. (1989) Mitochondial respiration can support NO_3^- and NO_2^- reduction during photosynthesis. Plant Physiol. **89**:409–415.

Weger, H.G., Herzig, R., Falkowski, P.G., and Turpin, D.H. (1989) Respiratory losses in the light in a marine diatom: measurements by short-term mass spectrometry. Limnol. Oceanogr. **34**:1153–1161.

Weinberger, D., and Porter, J.W. (1953) Incorporation of tritium oxide into growing *Chlorella pyrenoidosa*. Science **117**:636–638.

Weis, D., and Brown, A.H. (1959) Kinetic relationships between photosynthesis and respiration in the algal flagellate, *Ochromonas malhamensis*. Plant Physiol. **34**:235–239.

Weis, E., and Berry, J. (1987) Quantum efficiency of photosystem 2 in relation to energy-dependent quenching of chlorophyll fluorescence. Biochim. Biophys. Acta **849**:198–208.

Weis, E., and Lechtenberg, D. (1989) Fluorescence analysis during steady-state photosynthesis. Phil. Trans. R. Soc. Lond. **B323**:253–268.

Weiss, R.F., and Craig, H. (1973) Precise shipboard determination of dissolved nitrogen, oxygen, argon and total inorganic carbon by gas chromatography. Deep-Sea Res. **20**:291–303.

Welschmeyer, N.A., and Lorenzen, C.J. (1981) Chlorophyll-specific photosynthesis and quantum efficiency at subsaturating light intensities. J. Phycol. **17**:283–293.

Welschmeyer, N.A., and Lorenzen, C.J. (1984) Carbon-14 labeling of phytoplankton carbon and chlorophyll *a*: determination of specific growth rates. Limnol. Oceanogr. **29**:135–145.

Whitford, L.A., and Schumacher, G.J. (1961) Effect of current on mineral uptake and respiration by a fresh-water alga. Limnol. Oceanogr. **6**:423–425.

Whitledge, T.E., and Wirick, C.D. (1986) Development of a moored *in situ* fluorometer for phytoplankton studies. *In* Tidal Mixing and Plankton Dynamics, J. Bowman, M. Yentsch and W.t. Peterson (Eds.), Springer-Verlag, Berlin, pp. 449–462.

Wilhelm, C., Krämer, P., and Lenartz-Weiler, I. (1989) The energy distribution between photosystems and light-induced changes in the stoichiometry of System I and II reaction centers in the chlorophyll *a*-containing alga *Mantoniella squamata* (Prasinophyceae). Photosynth. Res. **20**:221–233.

Williams, P.J.LeB. and Jenkinson, N.W. (1982) A transportable microprocessor-controlled Winkler titration suitable for field and shipboard use. Limnol. Oceanor. **27**:576–584.

Williams, P.J.LeB., Raine, R.C.T., and Bryan, J.R. (1979) Agreement between the ^{14}C and oxygen methods of measuring phytoplankton production: reassessment of the photosynthetic quotient. Oceanol. Acta **2**:411–416.

Williams, P.J.LeB., Heineman, K.R., Marra, J., and Purdie, D.A. (1983) Comparison of ^{14}C and O_2 measurements of phytoplankton production in oligotrophic waters. Nature **305**:49–50.

Winkler, L.W. (1888) Die bestimmung des im wasser gelösten sauerstoffes. Chem. Ber. **21**:2843–2855.

Withrow, R.B., and Withrow, A.P. (1956) *In* Radiation Biology, A. Hollaender (Ed.), Vol. 3, McGraw-Hill, New York, pp. 125–258.

Wood, E.J.F. (1955) Fluorescent microscopy in marine microbiology. J. Cons. Int. Explor. Mer. **21**:6–7.

Woodward, F.I., and Sheehy, J.E. (1983) Principles and Measurements in Environmental Biology. Butterworths, London.

Wright, S.W., and Jeffrey, S.W. (1987) Fucoxanthin pigment markers of marine phytoplankton analyzed by HPLC and HPTLC. Mar. Ecol. Prog. Ser. **38**:259–266.

Wyatt, P.J., and Jackson, C. (1989) Discrimination of phytoplankton via light-scattering properties. Limnol. Oceanogr. **34**:96–112.

Wyman, M., Gregory, R.P.F., and Carr, N.G. (1985) Novel role for phycoerythrin in a marine cyanobacterium, *Synechococcus* strain DC2. Science **230**:818–820.

Yan, X-H., Schubel, J.R., and Pritchard, D.W. (1991) Oceanic upper mixed layer depth determinations by the use of satellite data. Remote Sensing Environ. (in press).

Yentsch, C.M. (1960) The influence of phytoplankton pigments on the color of sea water. Deep-Sea Res. **7**:1–9.

Yentsch, C.M., Yentsch, C.S., and Strube, L.R. (1977) Variations in ammonium enhancement, an indication of nitrogen deficiency in New England coastal phytoplankton populations. J. Mar. Res. **35**:537–555.

Yentsch, C.M. Horan, P.K., Muirhead, K., Dortch, Q., Haugen, E., Legendre, L., Murphy, L.S., Perry, M.J., Phinney, D.A., Pomponi, S.A., Spinard, R.W., Wood, M., Yentsch, C.S., and Zahuranec, B.J. (1983) Flow cytometry and cell sorting: a technique for analysis and sorting of aquatic particles. Limnol. Oceanogr. **28**:1275–1280.

Yentsch, C.S. (1962) Measurement of visible light absorption by particulate matter in the ocean. Limnol. Oceanogr. **7**:207–217.

Yentsch, C.S., and Menzel, D.W. (1963) A method for the determination of phytoplankton chlorophyll and phaeophytin by fluorescence. Deep-Sea Res. **10**:221–231.

Yentsch, C.S., and Yentsch, C.M. (1979) Fluorescence spectral signatures: the characterization of phytoplankton populations by the use of excitation and emission spectra. J. Mar. Res. **37**:471–483.

Yentsch, C.S., and Yentsch, C.M. (1984) Emergence of optical instrumentation for measuring biological properties. Oceanogr. Mar. Biol. Ann. Rev. **22**:55–98.

Yoder, J.A. (1979) A comparison between cell division rate of natural populations of the marine diatom *Skeletonema costatum* (Greville) Cleve grown in dialysis culture and that predicted from a mathematical model. Limnol. Oceanogr. **24**:97–106.

Yoder, J.A., Atkinson, L.P., Bishop, S.S., Blanton, J.O., Lee, T.N., and Pietrafesa, L.J. (1985) Phytoplankton dynamics within Gulf Stream intrusions on the Southeastern United States continental shelf during summer 1981. Cont. Shelf Res. **4**:611–635.

Zafiriou, O.C., Joussot-Dubien, J., Zepp, R.G., and Zika, R. (1984) Photochemistry of natural waters. Environ. Sci. Technol. **18**:358A–371A.

Zehr, J.P., Falkowski, P.G., Fowler, J., and Capone, D.G. (1988) Coupling between ammonium uptake and incorporation in a marine diatom: experiments with short-lived radioactive ^{13}N. Limnol. Oceanogr. **33**:518–527.

Zepp, R.G., Wolfe, N.L., Baughman, G.L., and Hollis, R.C. (1977) Singlet oxygen in natural waters. Nature **267**:421–423.

Zika, R.G., Moffett, J.W., Petasne, R.G., Cooper, W.J., and Saltzman, E.S. (1985) Spatial and temporal variations of hydrogen peroxide in Gulf of Mexico waters. Geochimica et Cosmochimica Acta **49**:1173–1184.

Index